PUBLIKATIONEN
DER BAYERISCHEN AMERIKA-AKADEMIE
Band 8

PUBLICATIONS
OF THE BAVARIAN AMERICAN ACADEMY
Volume 8

SERIES EDITOR
Board of Directors of the Bavarian American Academy

NOTE ON THE EDITORS

Klaus Benesch is Professor of English and American Studies
at the University of Munich
and Director of the Bavarian American Academy (Munich).

Meike Zwingenberger is the Executive Director
of the Bavarian American Academy (Munich)
and an adjunct Professor of American Studies at the University Munich.

Scientific Cultures – Technological Challenges

A Transatlantic Perspective

Edited by

KLAUS BENESCH
MEIKE ZWINGENBERGER

Universitätsverlag
WINTER
Heidelberg

Bibliografische Information der Deutschen Nationalbibliothek
Die Deutsche Nationalbibliothek verzeichnet diese Publikation in der Deutschen Nationalbibliografie; detaillierte bibliografische Daten sind im Internet über *http://dnb.d-nb.de* abrufbar.

UMSCHLAGBILD

Charlie Gunn, *Hyperbolic Space Tiled with Dodecahedron*
(Detail Turned)

ISBN 978-3-8253-5580-7

Imprimé en Allemagne · Printed in Germany
Gesamtherstellung: Memminger MedienCentrum, 87700 Memmingen

Gedruckt auf umweltfreundlichem, chlorfrei gebleichtem und alterungsbeständigem Papier

Den Verlag erreichen Sie im Internet unter:
www.winter-verlag-hd.de

Preface

Collections of critical essays are indebted—by their very nature—to the efforts and contributions of more than merely one author or one editor. Since many of the essays presented here evolved from the 2007 annual conference of the Bavarian American Academy (BAA) our gratitude goes out, first of all, to those who had either helped to organize or participated in the event as speakers, respondents, and critical discussants. Without the unflagging support of both the members of the Bavarian American Academy and the staff of the 'Amerika-Haus' in downtown Munich, where we convened, the conference would have been a much less invigorating and pleasant affair. That participants unanimously praised the warm and invigorating ambiance of the three-day conference attests to the outstanding professional commitment of those who conjoined to make this a successful event. Special thanks go to Patricia Guy, then Public Affairs Officer at the American Consulate in Munich, who supported us in more than merely financial ways, and Prof. Helmuth Trischler from the German Museum of Science and Technology (Deutsches Museum) for his participation as speaker and his generous invitation to the museum's premises.

More specifically, the editors want to thank Alondra Nelson (Yale University), Denise E. Pilato (Eastern Michigan University), and Rebecca Slayton (Stanford University), all of who could not participate in the original conference, but agreed to contribute extended original essays from their respective fields of research in science studies. Jasmin Falk, the BAA's office manager, and Daniele Franc, a student intern at the BAA, were crucial in preparing final drafts of the manuscripts and in handling copyright issues and illustrations. Last but not least we are grateful for the ongoing financial support of the Bavarian Ministry of Sciences, Research and the Arts without which neither the preceding conference nor the present collection of essays could have materialized.

Klaus Benesch and Meike Zwingenberger, the editors

Table of Contents

Science, Technology and the Literary Imagination

Technoscience and its Publics: Theories and Practices

Diverging Cultures, Competing Truths?

Science, Technology, and the Humanities: An Introduction

Klaus Benesch

> *There is something fascinating about science. One gets such wholesale returns of conjecture out of such a trifling investment of fact.*
>
> Mark Twain, *Old Times On the Mississippi* 407-8

> *Laws of nature are human inventions, like ghosts. Laws of logic, of mathematics are also human inventions, like ghosts. [...] The world has no existence whatsoever outside the human imagination. [...] It's run by ghosts. We see what we see because these ghosts show it to us, ghosts of Moses and Christ and the Buddha, and Plato, and Descartes, and Rousseau and Jefferson and Lincoln, on and on and on. Isaac Newton is a very good ghost. One of the best. Your common sense is nothing more than the voices of thousands and thousands of these ghosts from the past. Ghosts and more ghosts. Ghosts trying to find their place among the living.*
>
> Robert M. Pirsig, *Zen and the Art of Motorcycle Maintenance* 42

Given their obvious differences with respect to research goals, methodologies, and, most importantly, academic self-idealization the sciences and the humanities (particularly, the latest wave of theory-based literary and cultural studies) are often described as ontologically opposed modes of intellectual inquiry. While we associate the former with the search for truth through experimentation and the analysis of 'real,' i.e. physical matter, the latter are seen by many as interested at best in lofty ideas, aesthetics, and morality and, at worst, in ideology and the representation of unrestrained subjectivity. What is more, academic professionals in both camps appear to be equally divided, as writer/scientist C. P. Snow noted half a century ago, by "a gulf of mutual incomprehension—sometimes hostility and dislike—but most of all lack of understanding" (10).[1] Though still widespread within the scientific

[1] In his 1959 Rede lecture at Cambridge University the English novelist and physicist Charles Percy Snow claimed that there is a rift throughout contemporary Western intellectual life, "a gulf of mutual incomprehension" between writers and scientists, marked by "sometimes (par-

field itself, the view that the so-called 'hard' sciences, on the one hand, and the humanities, on the other, represent two separate, unbridgeable cultures of learning needs to be amended. Its implied bifurcation of scientific standards, values, and methods clearly falls short of capturing the complex relationship between science, technology, and the humanities in technologically advanced, posthuman societies. While both sides are involved in an increasingly fierce competition for public funding and recognition (what some observers have called the 'science wars'[2]), the progress made in computer science, robotics, genetic engineering and other cutting-edge disciplines raise new, far-reaching questions about the future of mankind as envisioned by scientists on both sides of the academic divide.

Political issues such as the dispute over evolution which turns on the right of religious groups to tamper with *what* and *how* science is taught at schools and universities also question the alleged authority of scientists and, eventually, add to the erosion of scientific standards. The lesson to be learned here is that the public understanding and social functions of the two cultures, to use Snow's notorious phrasing, are neither eternally fixed nor, as some 'fundamentalist' voices make believe, mutually exclusive. If we look at the history of the "two-cultures" debate, we find both an anxiety of the scientist about the waning of academic standards in the humanities and the humanist's outburst against the leveling of ethics and moral rules in the laboratory (just note the ongoing controversy about nuclear testing or genetic manipulation). What we rarely see, however, is a clear commitment to a third field, namely, the study of science itself. As historian of science Mario Biagioli points out, "the fact that science is a well-delineated and established enterprise seems to have two opposite effects on science studies: it allows the field to be simultaneously unified (in terms of its object of study) and strongly disunified (in terms of its methodologies, research questions, and institutional locations)" (XI). If it is true that the sciences are among the most convincing tools, as sociologist Bruno Latour argued, "to persuade others of who they are and what they should want" (259) then a concomitant investigation of the cultural and ideological underpinnings of scientific discourse, including its descriptive and normative registers, its proposed aims, research agendas and objects of study, seems in order. Put another way, only through a comparative analysis of various scientific cultures and the social and historical contexts in which scientists operate can we hope to even-

ticularly among the young) hostility and dislike, but most of all lack of understanding." Snow's famous lecture first appeared as "The Two Cultures" in *New Statesman*, 6 (October 1956). An expanded version, from which I quote here, can be found in: *The Two Cultures: And a Second Look* (1963). For a critical discussion of Snow's controversial thesis, see Joseph W. Slade and Judith Yaross Lee (eds.), *Beyond the Two Cultures: Essays on Science, Technology, and Literature* (1990).

2 For a number of different angles on America's 'Science Wars,' see Parson (2003) and Gross (1996).

tually overcome the "gulf of mutual incomprehension" and "lack of understanding" that Snow has identified as a major driving force behind the increasing tensions among academic communities.

To sketch out the often conflicted relations between science, technology and the humanities I shall begin by discussing two interesting cases of academic 'forgery': first, the *Codex Seraphinianus*, a fine, astonishingly complex piece of art written during the late 1970's by Italian graphic designer and architect Luigi Serafini; and, secondly, a fabricated poststructuralist critique of quantum theory by Alan Sokal, a professor of mathematical physics at New York University, which marked the highpoint of the so-called 'science wars' of the 1990's in the US and became widely known as 'Sokal's hoax.' If both their authors' intention and the formal means used to evoke a sense of authenticity differ tremendously, both texts nevertheless conjoin to shed new light on the allegedly "unbridgeable gap" between the two major academic cultures in the West, the humanities and the sciences or, in the case of Serafini and Sokal, art and physics.

Serafini's *Codex Seraphinianus* is a lavishly illustrated book that purports to be an encyclopedia in the style of Diderot's and d'Alembert's. As an elaborate parody of the real world and, more specifically, our desire to control this world by translating it into objectifying systems, classes, and groups of supposedly related phenomena it creates its own imaginary universe. Written in a florid, incomprehensible script and copiously illustrated with watercolor paintings the *Codex* is an attempt to evoke a consistent alternate world (albeit entirely invented) by copying both the writing style and categorical systematization of Western science. It is divided into eleven sections on diverse subjects such as plants, animals, ethnography, machinery, architecture, forms of numbering, cards, food culture, scenes of exotic festivals and even diagrams of plumbing – each partitioned into two parts with its own table of contents and, apparently, an intricate but equally unfathomable system of pagination. If many of the numerous illustrations appear to be parodies of items we find in the 'real' world, others are entirely made-up. They thus often evoke a sense of surreal magic and beauty (practically all are brightly colored and rich in detail) comparable to the experience of seeing a painting by Salvador Dali or Maurits C. Escher. Moreover, the writing system of the *Codex* seems to ape Western-style forms of writing: from left-to-right, in rows or columns, and based on an alphabet with uppercase and lowercase letters, of which some also double as numerals. Though the numbering system has apparently been deciphered, the language of the *Codex* still defies even the most attentive scrutiny and, after some twenty years of linguistic analysis, remains basically incomprehensible.[3]

[3] The Serafini's system used for numbering pages has recently been decoded by Bulgarian linguist Ivan Derzhanski (among others). See "*Codex Seraphinianus*: Some Observations" <http://www.math.bas.bg/~iad/serafin.html>.

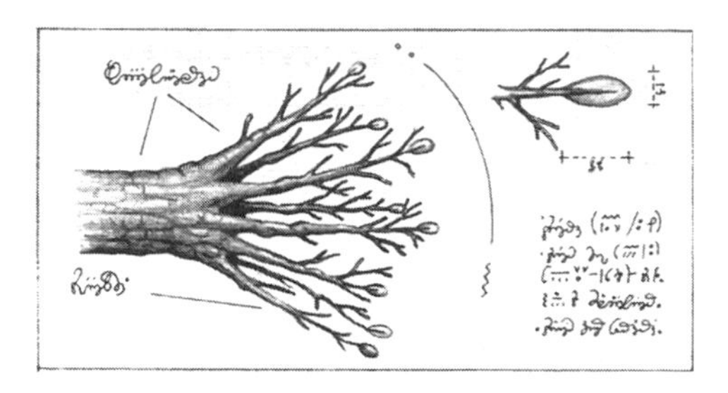

Treepaddles

Given its lavish and elaborate graphics, the *Codex Seraphinianus* can be read, for one, as pure art work, a rare and expensive specimen of postmodern hybrid art appreciated mainly by collectors and connoisseurs.[4] Yet for another, it may also be taken as a shrewd commentary on the scientific production of meaning through complex taxonomic apparatuses based on formal categorization and linguistic systematization. It's often grotesque content notwithstanding, the *Codex* is disturbing precisely because it purports to share—rather than dismiss—our sense of the real. Playing on the reader's recognition of specific modes of scientific representation (as in the writings of the French Encyclopedists or the natural histories of Linneaus, Alexander von Humboldt and Charles Darwin) Serafini's imagined encyclopedia is 'real' to the point of being truthful to even the most trifling details of science writing. Rather than critiquing a specific methodology, the selection of case studies, or the set up of experiments, Serafini's staggering artwork turns entirely on the mythopoeic potential of the formal rules of all science, its variegated systems of representation, classification, and taxonomy. By creating a mood comparable to the mood provoked by reading Darwin's *Origin of Species* or any other natural history classic, *Codex Seraphinianus* verges on the uncanny in that it allows the reader/viewer to navigate a scientific universe whose object, i.e. the world of nature and culture as we know it, has transmogrified into an undecipherable, alien Other.

As Douglas Hofstadter rightly remarks, Serafini relies "largely on style to convey content" (229). Parodying the formal conventions of scientific discourse his *Codex Seraphinianus* sheds light on the role of language in the process of formulating and disseminating specific scientific truths. The reader's disposition to accept a particular text as scientific depends to a great extent on its formal credibility rather than its inherent truth value. Both science and technology constantly gener-

[4] The original edition, which was issued in two volumes, has been out of print for many years and has now become a rare collector's item. Newer, single-volume editions such as the 2006 Italian edition, which features a preface by the 'author' and some added illustrations, are now available at relatively inexpensive rates but, given the special character of the book, have never garnered a large audience.

ate texts, graphic displays, and even moving pictures (as in endoscopic medicine or astrophysics); contrary to ordinary texts, however, scientific texts generally posit themselves as non-texts or representations of unequivocal, factual truths. By the same token, scientific authors do not 'possess' the ideas conveyed by a book or a journal article in the same way literary authors have copyright to the ideas conveyed by a novel. "According to definitions of intellectual property," as Mario Biagioli argues, "a scientist qua scientist is, literally, a nonauthor" (14). What Serafini's *Codex* thus makes apparent is precisely the crucial role of authorship in scientific discourse, of the style and form of presentation as essential ingredients for the construction of scientific authority. Far from being merely documentations of a particular research project, the texts of science are frequently aesthetic and profoundly cultural. They do not write *themselves* but are written *by* scientists who, more often than not, seek to avoid the conflicted issue of the social condition of their profession and instead postulate a view of science based on objective, disinterested truth, a pure universe unaffected by the vagaries of history, society, and culture.

This latter assumption has been at the core of a widely publicized controversy about a sham poststructural critique of quantum theory by New York physicist Alan Sokal which he submitted to the influential American cultural studies journal, *Social Text*, to see if the editors would accept it. The article was eventually published in a special issue of *Social Text* in 1996, devoted, as Sokal writes in the later, book-length publication *Fashionable Nonsense*, "to rebutting criticisms leveled against postmodernism and social constructivism by several distinguished scientists."[5] Contrary to previous academic hoaxes, this one served a public purpose, namely to attract attention to what Sokal saw as the declining standards of scientific investigation in the humanities. He therefore rapidly disclosed his prank in another journal, *Lingua Franca*, arguing that he had sprinkled his sham article with deliberate nonsense and that it was only accepted because "(a) it sounded good and (b) it flattered the editors' ideological preconceptions."[6] Consequently, Sokal's prank triggered a staggering number of responses and commentary, many of which pivoted on the more general contention that the humanities or so-called 'soft' sciences had degenerated into obscure, jargonized babble and that, in the process, it

[5] In an Introduction to *Fashionable Nonsense: Postmodern Intellectuals' Abuse of Science*, a book Sokal co-authored with French scientist Jean Bricmont (originally published in French as *Impostures intellectuelles*) 1-2. By assembling numerous quotations from a wide range of (then) contemporary French theorists and intellectuals such as Gilles Deleuze, Jacques Derrida, Jean-Francois Lyotard, Jean Baudrillard or Michel Serres, Sokal and his collaborator wanted to offer further, indelible proof that postmodernism (to their minds but a "nebulous Zeitgeist") has continuously abused both concepts and the terminology of mathematics and physics ("Introduction" 3).

[6] Quoted in Weinberg, "Sokal's Hoax" 11.

had given up entirely on the scientific objective of conveying accurate and, more importantly, comprehensible information.

Sokal HoaX

In a lengthy review of the controversy and its methodological implications nobel prize winner and physicist Steven Weinberg argued, for example, that within the limits of a technical language (such as the language of mathematics) scientists tend to be as clear as possible and would usually avoid "to confuse obscurity with profundity": "It never was true that only a dozen people could understand Einstein's papers on general relativity, but if it had been true, it would have been a failure of Einstein's, not a mark of his brilliance" (11). Put another way, academic excellence may well be a contested issue, yet it can never be a matter of rhetoric or obscure language alone. To take his point one step further, Weinberg actually claims that Sokal—while trying to imitate the humanities' knack for sheer jargon—was still writing in a relatively sober, straightforward prose. Unable to relish any kind of academic "babble" that has become fashionable with scholars in the humanities, Sokal would only deal in concepts and their proper meaning. In fact, as a physicist, he may have found it difficult "to write unclearly" (11). Rather than indulging the smug, self-serving style of, say, literary and cultural critics, he offers instead a compilation of quotes from deconstructionists such as Derrida who, according to Weinberg, blatantly distorted the concepts he borrowed from mathematics and physics.

One of Sokal's running jokes throughout the essay has to do with the word 'linear' and how it is used in the sciences and, vice versa, blatantly misused in the humanities. This word, as Weinberg points out,

> has a precise mathematical meaning, arising from the fact that certain mathematical relationships are represented graphically by a straight line. But for some postmodern intellectuals, "linear" has come to mean unimaginative and

old fashioned, while "nonlinear" is understood to be somehow perceptive and avant-garde. (11)

As two of the principal editors of *Social Text*, Bruce Robbins and Andrew Ross, explain in their defense (published in the following issue of *Lingua Franca*) and later in an open letter to *The New York Times*, Sokal may well have gotten it all wrong and thus attacked the journal on largely false grounds.[7] The glaringly Quixotean nature of Sokal's hoax notwithstanding, Weinberg's subsequent commentary reveals a more general misunderstanding about the methodology and discursive practices of the humanities. This is not the place to discuss the flimsiness of his contention—shared by many scientists, including Sokal and his collaborators—that science is a cumulative project and therefore allows historians to make definitive judgments about its success or failure, regardless of social, cultural, and political factors.[8] It should suffice here to recall George Santayana's caveat vis-à-vis the staggering positivism of late nineteenth-century America, namely, that "the ideal of rationality is itself as arbitrary . . . as any other ideal" (24). Not to mention the long list of modern critics and philosophers who repeatedly questioned the objective nature of scientific knowledge, from Nietzsche to Arthur Koestler, Paul Feyerabend, Thomas Kuhn, Bruno Latour or Richard Rorty (some of who Weinberg acknowledges as adversaries to his own belief that the laws of physics correspond to an objective reality).[9] However, in view of how knowledge is generated within different scientific cultures and, consequently, the possibility for mutual understanding and cross-fertilization among them (call it *interdisciplinarity*), Weinberg's above-quoted criticism of postmodernists' adoption of the mathematical concept of 'linearity' takes on particular importance. Its underlying notion that the success of science depends on an intra-disciplinary agreement about the precise

[7] As it's editors have repeatedly claimed, *Social Text*, for one, is not a refereed, professional science journal, but an originally self-published platform for political opinion and cultural analysis (though Duke University Press adopted its publication shortly before the appearance of Sokal's essay) that straddles the two fields of the independent left and the academic domain. And, for another, Sokal's piece has been included in the special issue on Science Studies precisely because of its inherent "queerness," as a "contribution from someone unknown to the field whose views, however peculiar, might still be thought relevant to the debate." Bruce Robbins and Andrew Ross, "Mystery Science Theater" 55; see also *The New York Times*, May 23 1996: 28.

[8] "If scientists are talking about something real, then what they say is either true or false. If it is true, then how can it depend on the social environment of the scientist?" ("Sokal's Hoax" 14).

[9] Nietzsche's idiosyncratic conception of a 'gay' science evolves from a similar effort to question the predominance and 'truth' value of scientific rationalism: "If we had not welcomed the arts and invented this kind of cult of the untrue, then the realization of general untruth and mendaciousness that now comes to us through science—the realization that delusion and error are conditions of human knowledge and sensation—would be utterly unbearable" (107).

meaning of concepts such as linearity, while in the humanities the same concepts have increasingly become hazy and malleable is, to say the least, problematic.

True, in the humanities ideas and concepts are more easily discarded because they do not figure as—more or less—accurate representations of physical reality (as in physics) but as expressions of the thoughts and ponderings of an individual. They may either subsume or repudiate entirely the ideas of previous thinkers and, simultaneously, serve to instigate further critical debates according to their proficiency to provide convincing interpretations of our social and individual existence. While each discipline agrees upon certain standards and formal rules to regulate intellectual debates, the production of meaning in the arts and humanities is to a large extent predicated on intellectual exchange, interaction, and the public performativity of ideas (i.e. their convincing presentation in a lecture, an essay, or monograph). Within the limits of their respective professional fields individual scholars are free to voice ideas totally idiosyncratic and/or revisionist. There is also a long history of attempts to cross boundaries and adopt themes and concepts from disciplines generally considered incompatible, thereby often changing not only the meaning of the adopted concept but also the assumptions and methodology genuine to the original research interest. If anything, Sokal's stunt thus turns out as beating a horse long dead. Since 'traveling' concepts change over the course of their academic journeys, the concept of linearity, to return to Sokal's own example, cannot but evoke different notions of connectedness, depending on whether it is used in physics, mathematics or philosophy.

To relate the meaning of concepts and methods to particular research contexts, what historian of science Donna Haraway has dubbed "situated knowledge," while at the same time opening up new avenues for interdisciplinary, border-crossing exchange and cooperation, is by no means a postmodern invention.[10] Just consider the following instances from the early nineteenth century, a period when the Romantic cult of creativity, on the hand, and the establishing of scientific positivism, on the other, conjoined to cement the rift between the *writing* of literature and the *doing* of science and technology. In an essay entitled "Of Persons One Would Wish to Have Seen," the British poet William Hazlitt writes in 1826, "I suppose the two first persons you would choose to see would be the two greatest names in English *literature*, Sir Isaac Newton and Mr. Locke" (32). Literature, in Hazlitt's understanding, was yet an implicitly transdisciplinary project that comprised the sciences as well as the investigation of ideas in philosophy. If Hazlitt's encompassing view of literary writing may be dismissed as both hopelessly romantic and

[10] In her frequently quoted essay, "Situated Knowledge: The Science Question in Feminism as a Site of Discourse on the Privilege of Partial Perspective," Haraway looks at the metaphors which science uses and how those metaphors help to determine (patriarchal) networks of power within the scientific field

vastly incompatible with modern scientific environments, nineteenth-century representatives of science were equally convinced that what connects the sciences to literature and the arts is perhaps more important than what separates them. In his widely read challenge of the Baconian method of induction (which emphasized the mechanical operations of experimentation and observation as the basic means of scientific research), the German chemist Justus von Liebig poignantly captures the mood among many of his scientific colleagues. Echoing the growing dissatisfaction with positivist materialism, von Liebig has come to conclude that "the mental faculty which constitutes the poet and the artist is the same as that whence discoveries and progress in science spring."[11] Both Hazlitt and von Liebig seem to believe that the sciences and the arts are intrinsically related rather than ontologically different, a closeness they see at work not only in the style and clarity of their best writing but in the mental faculties on which intellectual investigations in both fields depend.

If we replace science with the modern concept of technology, a replacement now frequently made by historians of science and technology, we often find an equally acute sense of relatedness and creative synergy among the two cultures.[12] As historian John Kasson noted, there has been a pronounced effort in nineteenth-century America to close the gap between the so-called *fine* and the *useful* arts (i.e. engineering and technology). The fact that artists such as painter Charles Willson Peale and sculptor Hiram Powers held a strong interest in machinery and that—vice versa—engineers such as Robert Fulton and Samuel F. B. Morse began their careers as painters has helped to substantiate, according to Kasson, "the comparisons that observers of technology frequently drew between machines and the fine arts and their contention that the two sprang from related imaginations" (146). The wedding of scientific rationality, on the one hand, and artistic imagination, on the other, also informs Edgar Allan Poe's tales of *ratiocination* and his holistic conception of the universe in the late, enigmatic prose poem *Eureka* (1848), subtitled "An Essay on the Material and Spiritual Universe." Poe considered both faculties as cornerstones of every intellectual activity and, in *Eureka*, he outlined his notion of capturing truth through language. If it's designation as 'prose poem' is somewhat misleading, *Eureka* underwrites Poe's intuitive conception of an universe in which the diverging forces of matter and spirit, of science and poetry, combine to endlessly rejuvenate the world by a process of expansion and contraction (like the

[11] Baron Justus von Liebig's essay, "Lord Bacon as Natural Philosopher", originally appeared in *Macmillan's* VIII (1863): 265. Quoted in Smith, *Fact and Feeling: Baconian Science and the Nineteenth-Century Literary Imagination* 35.

[12] Since modern technology always also involves science to a large degree the two terms have by now become interchangeable. See, among others, Stanley Aronowitz's collection of essays, *Technoscience and Cyberculture*. One classical study on this topic is Ellul, *The Technological Society* (1954) [originally published in 1954 as: *La Technique ou l'enjeu du siècle*].

Benjamin Franklin at his glass harmonica

beating of a heart). Given it's remarkable correspondence with recent cosmology, which has shown the universe to have evolved by constant expansion, Poe's *Eureka* stands as a striking example of poetic foresight, of reasoning by imagination.

In his magisterial study of creativity, *The Act of Creation* (1964), Arthur Koestler provides ample proof that it is indeed often intuition and the imagination—rather than pure reason or rationality—that drive scientific and technological progress. In keeping with Max Planck's remark in his autobiography that pioneer scientists must have "a vivid intuitive imagination for new ideas not generated by deduction, but by *artistically* creative imagination" Koestler points to an apparent paradox within the sciences: while scientific research is a branch of knowledge that "operates predominantly with abstract symbols, whose entire rationale and credo are objectivity, verifiability, logicality," it simultaneously depends "on mental processes which are subjective, irrational, and verifiable on after the event" (147). What is more, like a cartoonist, Koestler argues, a scientist represents processes by "a diagram, schema, map, or model . . . . He too is engaged in making models of phenomena in his particular medium, using a particular set of formulae, and concentrating on those aspects of reality, to the detriment of others, which are significant to him, or to the fashions and conventions of his time" (396). To articulate the beauty and relevance of a particular equation or an idea in philosophy, artists, scientists, and cartoonists alike employ techniques of "*selective emphasis*, exaggeration, and simplification" (397). As Koestler and numerous other commentators have shown, the fault lines between the two cultures of knowledge are thus neither fixed nor, as parochial voices on both sides of the divide make believe, unbridgeable. If the increasing rift between science, technology, and the humanities is by and large a modern phenomenon, we also find, as I hope to have shown, throughout the modern period efforts to cross borders and close the gaps between the two cultures.

In one of his late lectures, titled "The Question Concerning Technology" (1950), the German Philosopher Martin Heidegger claims that "the essence of the technical is by no means itself technical" (287). According to Heidegger it is impossible to explain the notion of technology by merely referring to its material manifestations, the world of tools and machines designed to alleviate and eventually supersede human labor. Rather technology, and by extension it's lofty double *science*, always also implies something bigger than itself, some force that transcends its immediate practical implications and cuts across all levels of our being in the world, in other words, the way in which we perceive that world as *real* and how we position ourselves within that perceived reality. For much of Western intellectual history to probe the *essence* of science and technology has been the uncontested domain of philosophy, arts, and the humanities. Sokal's sophomoric prank to the contrary, the recent intellectual trend toward transdisciplinary; 'culturalist' modes of academic research and education appear to narrow down rather than to widen the gap between different scientific cultures. Moreover, in conjunction with the staggering universal availability of highly specialized knowledge through electronic archives they may also help to increase our understanding of the cultural "situatedness" of all knowledge, the fact that rationality, to retool Santayana's phrasing, is indeed but an ideal and therefore as arbitrary and ambiguous as any other ideal.[13]

[13] For recent 'culturalist' approaches (particularly in science education), see Cobern and Loving, "Defining 'Science' in a Multicultural World: Implications for Science Education."

Works Cited

Aronowitz, Stanley (ed.) *Technoscience and Cyberculture*. New York: Routledge, 1995.

Biagioli, Mario. "Introduction." *The Science Studies Reader*. Ed. Mario Biagioli. New York: Routledge, 1999. XI-XVIII.

---. "Aporias of Scientific Authorship: Credit and Responsibility in Contemporary Biomedicine." *The Science Studies Reader*. Ed. Mario Biagioli. New York: Routledge, 1999. 12-30.

Cobern, W. W. and C. C. Loving. "Defining 'Science' in a Multicultural World: Implications for Science Education." *Science Education* 85 (2001): 50-67.

Ellul, Jacques. *The Technological Society*. Trans. John Wilkinson. New York: Vintage, 1964.Haraway, Donna. "Situated Knowledge: The Science Question in Feminism as a Site of Discourse on the Privilege of Partial Perspective." *Feminist Studies* 14.3 (1988): 575-99.

Hazlitt, William. "Of Persons One Would Wish to Have Seen." *New Monthly Magazine* (Jan. 1826): 32-41.

Heidegger, Martin. *Basic Writings*. Ed. David Farrell Krell. San Francisco: Harper, 1977.

Hofstadter, Douglas R. *Metamagical Themas: Questing for the essence of Mind and Pattern*. New York: Basic Books, 1985.

Koestler, Arthur. *The Act of Creation*. New York: The Macmillan Company, 1964.

Latour, Bruno. "Give Me a Laboratory and I Will Raise the World." *The Science Studies Reader*. Ed. Mario Biagioli. New York: Routledge, 1999. 258-275.

Nietzsche, Friedrich. *The Gay Science*. Trans. by Walter Kaufmann. New York: Vintage, 1974.

Parson, Keith. *The Science Wars: Debating Scientific Knowledge and Technology*. Amherst, NY: Prometheus Books, 2003.

Pirsig, Robert M. *Zen and the Art of Motorcycle Maintenance* [1974]. New York: Quill William Morrow, 1979.

Poe, Edgar Allan. "Eureka: A Prose Poem" [1848]. *Poe: Poetry and Tales*. Ed. Patrick F. Quinn. New York: Library of America, 1984.

Robbins, Bruce and Andrew Ross. "Mystery Science Theater." *Lingua Franca* (July/August 1996): 54-57.

Ross, Andrew (ed.) *Science Wars*. Durham, SC: Duke University Press, 1996.

Santayana, George. *The Sense of Beauty: Being the Outlines of Aesthetic Theory*. New York: Modern Library, 1955.

Serafini, Luigi. *Codex Seraphinianus*. 2 vols. Milano: Franco Maria Ricci, 1981 [127 + 127 pp./108 + 128 plates].

Slade, Joseph W. and Judith Yaross Lee (eds.) *Beyond the Two Cultures: Essays on Science, Technology, and Literature*. Ames, IA: Iowa State University Press, 1990.

Smith, Jonathan. *Fact and Feeling: Baconian Science and the Nineteenth-Century Literary Imagination*. Madison, WI: The University of Wisconsin Press, 1994.

Snow, Charles Percy. *The Two Cultures: And a Second Look*. New York: Mentor, 1963. 9-26.

Sokal, Alan and Jean Bricmont. *Fashionable Nonsense: Postmodern Intellectuals' Abuse of Science*. New York: Picador, 1999.

Twain, Mark. "Old Times On the Mississippi" [1857-60]. In: *Classic American Autobiographies*. Ed. William L. Andrews. New York: Mentor, 1992, 329-412.

Weinberg, Steven. "Sokal's Hoax." *The New York Review*. August 8 (1996): 11-15.

Science, Technology, and the Literary Imagination

Cultures of Risk and the Aesthetic of Uncertainty

Ursula Heise

The question of the relation between science and culture has been a contentious one over the last half century. If C.P. Snow, in his well-known hypothesis concerning the "two cultures" in 1959, had expressed a sense that humanities scholars did not know nearly enough about the development of modern science, scientists in the late 1980s and early 1990s began to be concerned that humanists and social scientists were encroaching altogether too much on natural scientific territory through their often critical sociological, anthropological, historical and cultural analyses of scientific practices. As the "Science Wars" culminated with Paul Gross and Norman Leavitt's vigorous counter-attack against such critiques and mathematician Alan Sokal's hoax in the journal *Social Text*, fundamental questions about the relationship of scientific research to its social and cultural contexts were discussed at great length in the mass media of the United States, Britain, and France. What distinctive features of scientific research set it apart from other forms of knowledge ranging from magic to religion and poetry? To what extent are scientific theories and methodologies shaped by the historical and cultural presuppositions of their day? Is scientific knowledge "ethnoscientifically" rooted in the complex of cultures commonly referred to as "Western"? What parts of scientific knowledge can be claimed to be "objective," which ones not? Can scientific practices themselves be studied in a scholarly or scientific way, and if so, by what means? If scientific practices are inexorably shaped by cultural contexts, how could the analysis of such contexts be any less so? Even after the Science Wars transmuted into quieter forms of dialogue in the late 1990s, such questions about the nature of the "real" and the forms of knowledge by means of which it can be grasped continued to be debated, shading in the early 2000s into anxieties over how the critique of the power-political imbrications of scientific inquiry might itself have turned into a tool of established political power.[1]

[1] The best-known articulation of this concern is Latour's "Why Has Critique Run out of Steam?"

One area of research where such questions have been debated from the early 1980s to the present day, and in which they translate into very concrete questions of policy and politics, is that of risk theory and risk analysis. An interdisciplinary field of research that has grown rapidly from its beginnings in the late 1960s and early 1970s, risk theory seeks to investigate how individuals, social groups and entire societies define and assess the risks that surround them, including medical, technological, environmental, political and economic ones.[2] Some of the more theoretical work in the field also investigates the connection of certain kinds of risk scenarios to the development of modern society at large, with its particular social, institutional, economic and technological dimensions. In these contexts, questions about the relative importance of scientific or statistical facts, cultural assumptions, and individual values and heuristics loom large in how they shape what dangers individuals and communities acknowledge and which ones they ignore, which ones they consider most and least threatening, and which ones they invest with cultural meaning. Situated at the intersection of science and culture, the investigation of risk perceptions also has important—and to date insufficiently studied—aesthetic implications. The literary myths of Dr. Faustus and Dr. Frankenstein have obvious relevance for the way in which scientific information about the dangers of modern technology is culturally filtered in many contemporary societies; less obviously but no less crucially, narrative templates such as that of pastoral or apocalypse, and narrative choices about the relation of the narrator to the narrative material, about protagonists and antagonists, and beginnings and endings shape the way in which risks are perceived and communicated, as do particular metaphors and visual icons.[3]

In tracing this connection between scientific investigations of risk, the cultural component of risk perceptions, and their articulation in works of literature and art, I will focus particularly on environmental risks, using some of the frameworks and methods that have evolved in the emergent field of ecocriticism for the last decade and a half. After a brief introductory survey of theories of risk, the essay will examine three works from different moments of the ecological movement, Rachel Carson's environmental classic *Silent Spring* (1962), German novelist Christa

[2] In tracing of risk theory back to the late 1960s, I follow the accounts that most scholars currently involved in the field give of its emergence (cf. Löfstedt and Frewer's introduction to *The Earthscan Reader in Risk and Modern Society* (3), and Lupton, *Risk,* chs. 1-2). Risk theory has a much longer history as a part of decision theory in economics, however, and it is possible to date its first origins as far back as late 18th century and epidemiological studies of vaccinations.

[3] Andrew Ross, for example, has pointed to the recurrence of the oil-covered seabird as an icon of ecological crisis (166, 171-72), and many other critics have commented on the synecdochic function of large and "charismatic" animal species in many environmentalist arguments. For a comprehensive analysis of the Frankenstein story and its impact, see Turney's *Frankenstein's Footsteps*.

Wolf's novel *Störfall: Nachrichten eines Tages* [*Accident: A Day's News*] (1987) and film-maker Steven Matheson's experimental short film *Apple Grown in Wind Tunnel: Wind Speed 85 MPH* (2000) to explore how they position scientific in relation to other kinds of language in their portrayals of pervasively contaminated environments. Scientific language, as I will argue, functions as a disruption of the normalcy of everyday life and discourse in these works, on one hand, but on the other hand it also comes to signal societies whose survival has become inconceivable without science – "scientific cultures" in an encompassing sense of the phrase. In negotiating the intersection of scientific with other types of language, these works recur to different kinds of metaphors to communicate the urgency of ecological and technological risks, but in the process they also attempt to develop a new kind of aesthetic whose force derives from structures of metaphor or irony that persistently interrogate the relationship between disaster and normalcy. In seeking out new ways of representing environments characterized by pervasive ecological and technological dangers, these works situate themselves within and aesthetically shape contemporary cultures of risk.

1. Risk Theory and the Question of Culture

In 1984, the engineer Charles Perrow published his by now classic study of technological hazards in modern societies, *Normal Accidents: Living with High-Risk Technologies*. Industrialization, and particularly the new technologies introduced over the course of the twentieth century, Perrow argues, have given rise to risks that are qualitatively different from earlier large-scale hazards such as epidemics, natural disasters or military conflicts in that they arise from systems so complex that not even experts can understand the interactions of all their subsystems or predict their most dangerous malfunctionings. In often fascinating detail analyses that range from dams and mines to air traffic, nuclear power and biotechnology, Perrow profiles the nature of what he calls "system accidents," which result from failures—often so minor that they are initially not even noticed by the system operators—in independent but coupled subsystems that end up producing large-scale failures in the system as a whole. The risks that result from these interactions are quite literally unpredictable in that they cannot be forecast on the basis of analyzing the overall system's normal functioning or subsystem malfunctions. "If interactive complexity and tight coupling—system characteristics—inevitably will produce an accident, I believe we are justified in calling it a *normal accident*, or a *system accident*. The odd term *normal accident* is meant to signal that, given the system characteristics, multiple and unexpected interactions of failures are inevitable. This is an expression of an integral characteristic of the system, not a statement of frequency,"(5) Perrow explains. Improved designs or better operator training proves ineffective in preventing such accidents because the complexity of the technology itself will always defeat them. Complex and tightly coupled systems, as Perrow demonstrates strikingly in his analysis of the accident at the Three Mile Island nuclear power plant in 1979, generate the most serious risks modern society is exposed to, and some such technologies are in his view too dangerous to be allowed to continue to operate.

Perrow's concept of the "system accident" or "normal accident" has justly become famous for the way in which it points to technological systems so complex that disastrous failure becomes ultimately unavoidable, and to the ways in which such risks have turned into integral parts of modern societies. Only two years after the publication of Perrow's study, the German sociologist Ulrich Beck proposed the concept of the "risk society" in what has become an ever better known if also far more sweeping and speculative theorization of risk and its relation to culture at the turn of the third millennium. As an alternative to the notion of a "postmodern society," Beck conceived this term to capture a particular dimension of what he, along with other sociologists such as Anthony Giddens and Scott Lash, calls "reflexive modernization," a phase of development in which modernization processes no longer operate on premodern social structures but on those created by earlier waves of modernization.[4] Reflexive modernization, according to Beck, gives rise to new kinds of hazards generated by modernizing processes themselves, and some of these risks, such as global warming and the thinning of the ozone layer, are for the first time truly planetary in scope. These new risks, in Beck's most daring claim, affect existing socio-economic stratifications in such a way as to create an entirely new kind of social structure that contrasts with those earlier stages of modernity that were organized around the production and distribution of wealth: "In advanced modernity, the social production of *wealth* is systematically associated with the social production of *risks*. Accordingly, the distribution problems and conflicts of the scarcity society are superseded by the problems and conflicts that originate in the production, definition and distribution of techno-scientifically generated risks" (*Risikogesellschaft* 25).[5] "Poverty is hierarchical, smog is democratic," Beck sums up this argument in one of the most provocative passages from *Risikogesellschaft* (48).

Beck is, of course, not unaware that current patterns of exposure to a wide range of hazards replicate structures of material inequality. From the siting of toxic waste dumps to the most dangerous kinds of employment and the unsafest types of housing, it is often the poor and disenfranchised—both nationally and internationally—that bear the greatest burden of risk. "[T]here is a systematic 'force of attrac-

4 See the volume *Reflexive Modernization: Politics, Tradition and Aesthetics in the Modern Social Order*, co-authored by Beck, Giddens and Lash.

5 All translations from Beck's *Risikogesellschaft* are mine.

tion' between extreme poverty and extreme risks," Beck emphasizes (*Risikogesellschaft* 55), implying that the emergent structures of risk society for the moment overlap with the structures of the modernist scarcity society. The point of his argument, therefore, is not that the increased number and scope of modern risk scenarios have already overridden existing social inequalities, nor that they will lead to an egalitarian society. Instead, they will eventually lead to a rearticulation of inequalities on a different basis: "One thing is clear. Endemic uncertainty is what will mark the lifeworld and the basic existence of most people—including the apparently affluent middle classes—in the years that lie ahead" (*World Risk Society* 12).

The consequences of this momentous social shift manifest themselves both politically and culturally, in Beck's view. The pervasiveness of risk, as he points out, has already led to the emergence of a global and increasingly influential "subpolitics" conducted not only by governments and the transnational institutions they form part of, but also by NGOs and other groups outside conventional political structures. Culturally, as Beck has argued in one of his more recent works, *Der kosmopolitische Blick* (2004), transnational risk scenarios open up the possibility of new cultural communities tied together by their shared experience of risk and their attempts to manage and control the hazards that threaten them. The risk society, in this view, implies the potential for new kinds of cosmopolitan awareness that reach across existing national, economic and cultural borders to encompass all those affected by risk scenarios ranging from polluting industries to species extinction or climate change. Rather than portraying the gradual transition to the risk society as the deplorable end result of several centuries of modernization, therefore, Beck's sociological theory, especially in its more recent formulations, also seeks to highlight the positive socio-cultural innovations that might result from this historical shift.[6]

The British sociologist Anthony Giddens, some of whose work is clearly influenced by Beck, theorizes the connection between risk and culture at the turn of the third millennium in similar terms. One of the most fundamental characteristics of modernization processes, according to Giddens, is the "disembedding" of government, social institutions, and networks of exchange and expertise, that is, their shift from local contexts to regional, national and in some cases transnational ones. Even as such disembedding mechanisms create safety networks for large areas and populations, for example through steady and safe supplies of food, water and elec-

[6] Beck's theory of cosmopolitan consciousness and culture, however, cannot but strike one as simplistic compared to the much more sophisticated approaches to this issue that have been formulated in literary and cultural studies as well as in anthropology, philosophy and political science over the last decade. I give a more detailed critique of Beck's conceptualization of the connection between risk and culture in *Sense of Place and Sense of Planet: The Environmental Imagination of the Global*, ch. 4.

tricity, shared legal conventions, or insurance practices, they also give rise to unprecedented risks:

All disembedding mechanisms take things out of the hands of any specific individuals or groups . . . . Despite the high levels of security which globalised mechanisms can provide, the other side of the coin is that novel risks come into being: resources or services are no longer under local control and therefore cannot be locally refocused to meet unexpected contingencies, and there is a risk that the mechanism as a whole can falter, thus affecting everyone who characteristically makes use of it (*Consequences of Modernity* 126-127).

One of the most significant consequences of the emergence of such new risks is that they put to the test the social trust that is indispensable for the functioning of modern societies, most of whose institutions no longer rely on personal knowledge and face-to-face interaction. Risk scenarios, especially exceptionally serious or large-scale ones, endanger this infrastructure of trust, since modern networks of information and communication enable widespread awareness of a host of different risks, as well as awareness of the limits of expertise in dealing with them. In addition, modern societies typically do not offer their members easy ways of converting such limits of knowledge or management ability into the certainties of magical or religious knowledge (*Consequences of Modernity* 125). The change in the kind of risk scenarios that disembedding mechanisms create, in combination with the type of risk awareness they give rise to, leads Giddens to define late modernity as a "risk culture" (*Modernity and Self-Identity* 3), clearly a term that in some basic respects parallels Beck's concept of the "risk society."

Beck pinpoints the cultural consequences of such shifts in the dialectic of dependence and distrust that ties the public to the institutions that manage risk by discussing in detail the role of mass media, scientific experts and government agencies in creating or dissipating certain risk perceptions in the public sphere.[7] Implicitly at least, Beck here draws on a wide range of work that has been undertaken since the mid-1980s on the so-called "social amplification" and "social attenuation" of risk, that is, the role that formal as well as informal networks of information and communication, ranging from the media, universities, churches and schools all the way to families, circles of friends and Internet chat groups, play in the shaping of risk perceptions among the non-expert public (see Kasperson, and Kasperson et al.) But Beck, in typically provocative manner, goes one step further than most analysts of the social mediation of risk by postulating that the intensified awareness typical of the risk society in fact puts in question the underlying cultural logic of everyday life:

[7] See Wynne's "May the Sheep Safely Graze?" for a critique of Giddens' and Becks' theorizations of the public relationship with expert discourses and their institutions.

> In order to perceive risks as risks and to make them a reference point for one's own thought and action, one has to believe in fundamentally invisible causal connections between conditions that are often substantively, temporally and geographically far removed from each other, as well as in more or less speculative projections . . . . But that means: the invisible, more than that: that which as a matter of principle cannot be perceived, that which is only theoretically connected and calculated becomes . . . an unproblematic component of personal thought, perception, experience. The "experiential logic" of everyday thought is, so to speak, turned upside down. One no longer only induces general judgments from one's own experiences, but instead general knowledge that is not based on any experience becomes the determining center of one's own. Chemical formulae and reactions, invisible toxins, biological circuits and causal chains must dominate vision and thought to lead to active fighting against risks. In this sense, risk awareness is not based on "second-hand experience," but on "second-hand non-experience." Even more pointedly: Ultimately no one can know of risks if knowing means having consciously experienced them. (Risikogesellschaft 96)

Beck does not discuss in any detail by what means scientific data and forecasts become integral parts of everyday culture and thought. Social scientists studying risk perceptions and risk communication have paid far greater attention to this question in examining what factors go into the making of non-expert risk judgments, an issue that I will examine in more detail shortly. From the perspective of literary and cultural studies, the role of particular narrative templates—such as pastoral, apocalypse or the detective story—and of the metaphors by means of which novel and unknown risk scenarios are assimilated to known ones deserves close analysis so as to determine the cognitive shape a "risk culture" takes on in everyday life. As I will argue in later sections of this essay, works of literature and art with an environmental focus have long struggled with the question of what narrative and metaphoric patterns might be appropriate for conveying a sense of global ecological crisis, and have at the same time reflected on the ambiguous politics of rhetorical devices that familiarize the public with the risks it is exposed to, but at the same time threaten to normalize them to the point where they come to form part of everyday routine. As my analysis will show, the consequences of Beck's "second-hand non-experience" are therefore less straightforward than he makes them appear: the integration of scientific knowledge about current risk scenarios into everyday thought may be the foundation for social activism, but it can by the same token become the basis for the acceptance of risk as part of everyday routine. Implicitly or explicitly, it is this ambiguity that the works I will examine address through their verbal and visual strategies.

In the social sciences, the question of how risk perceptions arise and what shape they take in different social groups has been studied in great detail since the early 1970s. But when the engineer Chauncey Starr gave the initial impulse for this kind of investigation with his studies of the relationship between assessments of technological risks and social benefits in the late 1960s, the question of culture did not enter into his analysis in any significant way. At a time when the public in many Western countries adopted an at least partially skeptical attitude toward science and technology in a context of concern over the potential of nuclear war and environmental apocalypse, Starr sought to examine what trade-offs between technological benefit and risk were considered publicly acceptable, which ones not, and for what reasons. Over the course of the 1970s, sociologists and cognitive psychologists followed his lead by studying public perceptions of a multitude of risk scenarios ranging from earthquakes and car accidents to smoking, firearms, pesticides and nuclear power plants. As they developed what came to be called the "psychometric paradigm," these researchers focused on empirical data generated through surveys and examined what cognitive selection and filtering processes shape risk perceptions. As it turned out, the public at large quite often diverged from experts in science, engineering and statistics in its judgments; whereas such experts tended to compute risks on the basis of their magnitude and probability of occurrence, the public used different models in its assessments. Psychometric researchers found, among other things, that voluntarily incurred risks tended to be rated less dangerous than those to which individuals were exposed involuntarily; dangers imperceptible without special equipment were often rated more hazardous than directly observable ones, new risks more dangerous than old ones, unfamiliar ones larger than those already known. Similarly, risks whose effects are delayed tended to appear more threatening than those with immediate effects, those with controllable or non-fatal consequences less threatening than those with uncontrollable or lethal ones, and those with mostly local implications less dangerous than those with national or global ones. As statistical analysis reveals, these varied components of risk judgments sometimes combine to give rise to a particular "dread" factor that is associated with some risks but not others: nuclear radiation or cancer, for example, tend to be perceived with such dread whereas car accidents or heart disease are not (Fischhoff et al., *Acceptable Risk*, chs. 4-7; Slovic, "Perception of Risk;" Fischhoff, Slovic and Lichtenstein, "Lay Foibles and Expert Fables").

Clearly, then, experts and non-professionals use different evaluative procedures in their approaches to risk. Throughout most of the 1970s, the prevailing assumption was that the public's views of risk, insofar as they diverged from the "objective" assessment of the experts, needed to be "corrected" through improved risk communication. This perspective increasingly came to be challenged through the cultural turn risk analysis took over the course of the 1980s. In their controversial book *Risk and Culture* (1982), the anthropologist Mary Douglas and the sociologist Aaron Wildavsky proposed a different approach to the study of risk perceptions by suggesting that their underlying reason should not primarily be sought in a combination of the properties of the risk scenarios themselves with the cognitive biases and heuristics of the individuals assessing them, but in terms of more encompassing worldviews such individuals hold. Centrally, they argued, risk judgments are not made on a case-by-case basis, but can be predicted in terms of the kinds of social structures individuals are involved in and committed to. These risk perceptions, in turn, play a significant role in the maintenance and self-perpetuation of such social structures. While Douglas and Wildavsky's basically argued that all risk perceptions are culturally constructed in this way, their case analyses singled out the rapidly growing environmentalist movement to show that claims concerning ecological crisis had less to do with the state of nature than with the egalitarian social structures environmentalist groups tended to adopt.[8] Even though the risk assessments of government agencies or corporations should, according to Douglas and Wildavsky's theory, be amenable to a similar analysis in terms of underlying social forms of organization, their single-minded critique of environmentalists earned *Risk and Culture* a barrage of criticism from social scientists who objected to its analytical terms and from environmentalists who resisted its ideological bias.

Douglas and Wildavsky's approach (which Douglas herself revised in some of its details in subsequent publications) turned out to be difficult to translate into the operational terms of empirical social-scientific research, and were not always confirmed when such empirical testing was undertaken. But their work gave the initial impulse for varied kinds of research that investigate risk perceptions in their social, cultural and ideological contexts. Indeed, perhaps the most crucial contribution of Douglas and Wildavsky's brand of cultural theory may be the insight that risk perceptions emerge from particular cultural contexts without which the very notion of risk cannot even be conceived. Many of these cultural variables have by now been incorporated into psychometric approaches which are today no longer nearly as distinct from their cultural-theoretical counterparts as they were in the 1980s (cf. Slovic, "Trust" 402). An ample body of research now exists on divergences in risk perceptions between men and women, different racial and ethnic groups, and different nations and regions. The role of broadly understood social, cultural and po-

[8] Douglas and Wildavsky used a grid-group model to characterize four different types of worldview and social structure, with the concept of "group" refering to the range of social interactions and degree of social integration and the concept of "grid" describing the nature of these interactions in terms of hierachy, access to particular social roles, etc. (cf. Rayner 87): egalitarianism, individualism, hierarchism and fatalism. As Rayner points out, however, Douglas and Wildavsky quickly reduced this fourfold distinction to a simple dichotomy between social center and margin, "a traditional conflict of interests between a conservative industrial rationality and a radical rural-idealist opposition" (91).

litical perspectives and worldviews in shaping such perceptions is no longer a matter of contention. Important dimensions of risk perception such as trust in the institutions charged with managing risks and the social amplification and attenuation of risk mentioned earlier have been studied extensively; additional variables such as affect are under investigation.[9]

Studies of risk perceptions are continuously being conducted, and their empirical findings as well as the statistical methodologies and theoretical frameworks used to evaluate them are subject to vigorous debate. What has emerged clearly from several decades of research and theorization, however, is that risks cannot be investigated outside the socio-cultural contexts that shape them and endow them with meaning. Increasing public discussion of a wide range of hazards as well as the popularization of the risk concept itself over the last two decades, in addition, have led to a culture, in Western societies and around the globe, in which awareness of multiple risk scenarios has become so common a constituent of everyday consciousness that Giddens' notion of a "risk culture" is surely justified at least in this sense. For the writers and artists who engage with this awareness of a pervasive risk environment, especially those who do so with a sense of social engagement, the question arises as to how to negotiate imaginatively the tension between fear and routine, deadliness and banality, disaster and the everyday. The three works I will examine, responding to quite different historical moments between the 1960s and the present day, reflect a changing sense of this relationship between risk and routine. As Frederick Buell has persuasively shown, the expectation of future catastrophe that prevailed with regard to both nuclear conflagration and environmental collapse in the 1960s subsequently transmuted into an awareness of ongoing crisis in the present (*From Apocalypse to Way of Life* 177-208). Instead of anticipating disaster, F. Buell argues, most populations have learned to live with, and sometimes to accommodate to, a multitude of daily risk scenarios – in the cases he is concerned with, most centrally ecological risk scenarios. Carson's *Silent Spring*, Wolf's *Störfall* and Matheson's *Apple Grown in Wind Tunnel* translate these changing cultures of risk not just into their thematic concerns but also into their rhetorical and aesthetic strategies.

[9] The bibliography on these issues is vast. A few representative titles that refer to a good deal of additional work include Slovic, "Trust, Emotion, Sex, Politics, and Science: Surveying the Risk-Assessment Battlefield"; Finucane et al., "The Affect Heuristic in Judgments of Risks and Benefits"; Renn and Rohrmann, *Cross-Cultural Risk Perception*; Sjöberg, "Are Received Risk Perception Models Alive and Well?"; and "Principles of Risk Perception Applied to Gene Technology."

2. Science and Metaphor in the Risk Society

Texts and artworks that have sought to engage with environmental crises over the last forty years have persistently faced the challenge of how to negotiate the ambiguous relationship with the natural sciences that has characterized environmentalism during the same period. While environmentalists are aware that many of their claims regarding the state of ecosystems as well as human health centrally rely on scientific insight, scientific rationality has also often been cast as one of the root causes of ecological crisis in environmentalist thought.[10] This ambivalence also makes itself felt in the way in which texts and works of art with an environmentalist focus incorporate scientific information. At a practical level, they face the necessity of translating this information into terms that are intelligible to the general public, especially where as yet unknown risks are concerned: exposure to chemical toxins from pesticide use or imperfectly secured waste dumps constituted such an unfamiliar risk scenario in the 1960s, for example, while in the 1980s and 90s dangers associated with genetically modified organisms emerged into public discourse and required explanation. At a more theoretical level, environmentally oriented works often oscillate between the necessity of conveying such basic scientific information and their own reservations about the dominant role of scientific discourse in regulating the daily lives of people in contemporary society. Scientific discourse, in such texts, therefore enters into a complex dialectic with the texture of everyday life, disrupting its normal routines at the same time that it establishes the new normalcy of pervasive "riskscapes" through the recurrent deployment of comparisons and metaphors.[11]

Such structures of metaphoricity have quite often been pointed out in Rachel Carson's *Silent Spring* (1962), the classic text that is often credited with triggering the modern environmentalist movement, and which first popularized the "totalizing images of a world without refuge from toxic penetration" that subsequently became a characteristic feature of what ecocritic Lawrence Buell has called "toxic discourse" (*Writing for an Endangered World* 38). Focusing on the overuse of pesticides and herbicides in agriculture and households as a lethal danger to wildlife and a serious threat to human health, *Silent Spring* seeks to convey to a general audience the findings of scientific studies that had been carried out in the 1950s but had not received much public attention. Unlike two very similar books that were published at approximately the same time, Lewis Herber's (alias Murray Bookchin's) *Our Synthetic Environment* (1962) and Robert Rudd's *Pesticides and the Living Landscape* (1964), *Silent Spring* succeeded in bringing two central ideas

[10] I have explored this ambivalence in greater detail in the essays "Science and Ecocriticism," "Science, Technology and Postmodernism."

[11] The term *riskscape* is Susan Cutter's, as quoted in Deitering (200).

into the realm of public debate. In what has sometimes been referred to as a new "ecology of the human body" (Lear xvi), Carson pointed out that chemicals released into the environment do not progressively become more diluted as they travel from organism to organism, but on the contrary more concentrated as contaminated insects, for example, are eaten by fish who are in turn eaten by birds or mammals and so on up the food chain, leaving humans as the top predators absorbing the largest concentrations of toxins. She persistently emphasized the permeability of the human body to an environmental contamination from which it cannot be kept immune. This circulation of poisons through ecological systems in turn led her to indict pesticide use as an attack on the very "fabric" (197) or "web of life," against the "integrity of the natural world" itself (13), thereby creating the image of a totally contaminated world that Buell highlights.

Carson succeeded in conveying these ideas to a general public through the use of three well-studied strategies. The first of these is the use of pastoral and apocalyptic narrative templates in the best-known, and to many readers, only known part of *Silent Spring*, the so-called "Fable for Tomorrow" that introduces Carson's argument by portraying a small town in the American heartland overcome by a blight of cataclysmic proportions that threatens its very existence.[12] A second and more subtle procedure is her reappropriation of some of the rhetoric of McCarthyist anti-communism to portray chemical exposure as another sneaking, unseen and lethal danger to the well-being of Americans in such phrases as "a grim specter stalks the land" (3). Her third and perhaps most important strategy, however, associates the dangers of agricultural and household chemicals with the risk scenarios of nuclear war and radioactivity.[13] In contrast to chemical toxicity, the dangers of radioactivity were well-known due to Hiroshima, Nagasaki and the Cold War, and came accompanied by a whole host of images and stories, from mushroom clouds and deformed bodies all the way to the innumerable radioactive mutant films Hollywood churned out during the 1950s. Drawing on such popularly known risk scenarios, Carson warns of the dangers of genetic damage, for example, by pointing out that "[w]e are rightly appalled by the genetic effects of radiation; how then, can we be indifferent to the same effect in chemicals that we disseminate widely in our environment?" (37), and by insisting that where genetics is concerned, "the parallel between chemicals and radiation is exact and inescapable" (208). In the context of birth malformations, she urges that "it must not be overlooked that many chemicals are the partners of radiation, producing exactly the same effects" (205), and she

12 For some exemplary discussions (among many others that have appeared on Carson's rhetoric), see Buell, *The Environmental Imagination* 44 and 290-296; Garrard, *Ecocriticism* 94-96; Killingsworth and Palmer, "Millennial Ecology"; and Patrick, "Apocalyptic or Precautionary?"

13 Glotfelty discusses both the anti-communist and the nuclear rhetoric of *Silent Spring* in her essay, "Cold War, Silent Spring."

concludes: "in all the thousand million years [of evolution] no threat has struck so directly and so forcefully . . . as the mid-20th century threat of man-made radiation and man-made and man-disseminated chemicals" (210-211). Clearly, by persistently associating the vocabulary of nuclear threats with that of chemical toxicity, Carson meant to tap a culturally well-established pool of risk perceptions in which she could then float the scientific facts about the much less familiar danger. Far from icons of technological progress in the house and on the farm, she implies, such chemicals threaten the daily routines of domestic and rural life just as terminally as the atomic bomb.

When Beck wrote *Risk Society* approximately a quarter of a century later, the idea of a pervasively polluted environment had become well established in the public imagination. Lawrence Buell, indeed, calls him "the Rachel Carson of contemporary social theory" (*The Future of Environmental Criticism* 5). But while Carson described the normal operations of one kind of modern technology, Beck's account is also informed by the experience of Perrow-style "system accidents," the ones that are not supposed to happen in advanced technological systems but do anyway. In Europe especially, the explosion of a nuclear reactor in the Ukraine in April, 1986, became a shaping event for the cultural awareness of pervasive risk. The vast scale of contamination from the Chernobyl explosion – more than twenty countries and 400 million people were subject to fall-out from the accident – conveyed a Carsonian sense of a world everywhere contaminated and a technology whose consequences and side effects were liable to spin out of control at any time. Beck's preface to *Risk Society*, dated May 1986, refers explicitly to Chernobyl: "Die Rede von (industrieller) *Risikogesellschaft* . . . hat einen bitteren Beigeschmack von Wahrheit erhalten. Vieles, das im Schreiben noch argumentativ erkämpft wurde—die Nichtwahrnehmbarkeit der Gefahren, ihre Wissensabhängigkeit, ihre Übernationalität, die 'ökologische Enteignung,' der Umschlag von Normalität in Absurdität usw.—liest sich nach Tschernobyl wie eine platte Beschreibung der Gegenwart. Ach, wäre es die Beschwörung einer Zukunft geblieben, die es zu verhindern gilt!" [The discourse about an (industrial) *risk society* . . . has been given a bitter taste of truth. Much of the argument I still had to struggle for in the writing—the imperceptibility of hazards, their dependence on knowledge, their transnational character, 'ecological expropriation,' the switch from normalcy to absurdity, etc.—reads like a flat description of the present after Chernobyl. Oh, that it had remained the evocation of a future to be prevented!] (*Risikogesellschaft* 10-11).

The accident became a central occasion for writers and artists in both Eastern and Western Europe to reflect on the nature of modern risk scenarios. Christa Wolf, one of the best-known writers in what was then the German Democratic Republic, undertook such reflections in her controversial novel *Störfall: Nachrichten eines Tages* (1987) – controversial because, between 1988 and 1990, dozens of scientists and intellectuals debated *Störfall* in the pages of the scientific magazine *spectrum* and in public forums at the East German Academy of Arts. The novel describes a single day in the life of an ageing writer who lives in a small village in Mecklenburg, a day filled with an array of ordinary events and activities: preparing meals, gardening, talking to neighbors, riding a bike, making phone calls, reading a book, listening to the radio and watching TV. But these unremarkable pursuits contrast sharply with the invisible and life-threatening events that intrude upon the narrator's life and occupy her thoughts throughout the day. Her brother is undergoing brain surgery for a dangerous tumor, and for many hours she anxiously imagines the various steps of the operation and their possible consequences, until an afternoon phone call informs her of its successful outcome. During her wait, she follows the unfolding news about Chernobyl with increasing unease and outrage. Repeatedly, she questions in her mind what the cultural and perhaps even evolutionary origins might be of the fascination with technology and the disdain for nature and human life that brings about disasters such as Chernobyl. This criticism of nuclear technology and the scientific urge to know led scientists and engineers to interpret Wolf's novel as an unqualified rejection of nuclear energy, which they either criticized or endorsed. Literary scholars, by contrast, have often pointed out that the novel portrays both destructive and creative, pathogenic and therapeutic technologies through the juxtaposition of the Chernobyl disaster and the successful brain surgery (Brandes 107; Eysel 293; Hebel 43; Kaufmann 256; Magenau 346; Rey 375; Weiss 102).

From the interpretive perspective of risk analysis that I am proposing here, however, *Störfall* presents itself less as a binary view at contemporary technologies than as a portrait of different kinds of risk: known and unknown ones, voluntarily and involuntarily incurred ones, perceptible and imperceptible ones, local and regional ones.[14] In both the context of brain surgery and that of nuclear disaster, the narrator is confronted with scientific expert knowledge on whose implications for her everyday life she meditates throughout the day. But while the technical and medical details of the tumor operation seem to pose no significant obstacle for her understanding, the scientific language associated with the Chernobyl accident disorients her to the point where she surmises that it might entail a wholesale reorganization of different strata of language ranging from the colloquialisms of everyday conversation to poetic and religious rhetoric. The question that arises in this context is precisely one of comparison and metaphoricity – that is, the question of what other threats Chernobyl can meaningfully be compared to. It is in the end this question of comparison and comparability in the face of large-scale technological threats, I would argue, that underlies the bifurcation of the novel's themes as well as many of its other allusions to history and literature.

[14] I have proposed a related but somewhat differently slanted interpretation of *Störfall* in the context of other works about the Chernobyl accident and the way in which they negotiate local and global dimensions of risk in "Afterglow: Chernobyl and the Everyday."

In her portrayal of both medical and nuclear technologies, the narrator emphasizes the gap that separates non-professionals from experts when it comes to information about risk. "Ich gebe zu, ein wenig beunruhigt mich die Tatsache, daß diese Leute, wie alle Spezialisten, den heiligen Schauder vor den Abgründen ihres Faches nicht mit uns Laien teilen können; daß sie also, in unvermeidlicher Berufsroutine, deine und meine ehrfürchtige Scheu vor einem Eingriff in jene Sphäre verloren haben, in der beschlossen liegt, ob wir so oder anders sind" (68) [I will admit, the fact that these people, like all specialists, cannot share the holy terror of the yawning abyss of their particular field with us lay people is a bit disconcerting; in other words, they have lost, through unavoidable professional routine, your and my reverend fear of operating within that sphere (i.e. the brain) in which is decided whether we are one way or another (44)].[15] In spite of this gap, however, the narrator considers details of the brain surgery throughout the day with a fair amount of scientific detail, reflecting on where and how the doctors will open up her brother's skull, what parts of the brain they will come in contact with and which ones they will avoid, as well as what consequences for her brother's sensory and cognitive abilities the surgery, including possible mishaps, might have.[16] At one point, she reflects, for example: "sie finden jene Masse, die sich unter dem Mikroskop in Zellen auflösen würde: Neuronen; und sie könnten, bei starker Vergrößerung, jene 'Synapsen' benannten Verbindungen zwischen den Neuronen finden, deren Menge größer ist als die Gesamtzahl der Elementarteilchen im Universum" (62-63). [they are finding the mass which would dissolve into cells beneath the microscope: neurons; and they could, with strong magnification, find those connectors between the neurons called 'synapses,' which number greater than the total amount of primary particles in the universe (40)].

But if reflections such as these seem to indicate a fair amount of medical literacy on the narrator's part, she nevertheless often domesticates this scientific perspective by considering the banal materials, tools and procedures in the surgery that seem to link it to the activities of everyday life. In this vein, she wonders what

[15] Quotations from *Störfall* in English are taken from Schwarzbauer and Takvorian's translation, to which the parenthetical page references following English quotations from the novel refer. In double page references, the number before the slash refers to the German and the number after it to the English edition.

[16] "Deine Ärzte, nach dem säuberlich hakenförmigen Schnitt über der rechten Augenbraue, nachdem sie alle Blutgefäße sorgfältig abgeklemmt, die Kopfhaut nach beiden Seiten so weit wie möglich auseinandergezogen haben, die Hirnmasse beiseitegeschoben und in Folie verpackt, konzentrieren sich vermutlich darauf, dem Kern des Übels so nahe wie möglich zu kommen, ohne die Hypophyse zu verletzen," she thinks in the morning (32). [After the clean, hook-shaped incision above your right eyebrow, after all the blood vessels have been meticulously clamped, the scalp stretched apart as far as possible to both sides, the brain pushed aside and wrapped in foil, your doctors are presumably concentrating on getting as close to the root of evil as possible without damaging the pituitary gland (19)].

kinds of saw, drill and pump might be used to open up the skull and drain fluids (15, 57/7, 36), whether the instruments used inside the brain would be made of metal or not (33/20), and whether the thread used in surgery whereby her brother's life literally hangs is made of cat or sheep gut or a synthetic suturing material (19, 40/11, 25). More metaphorically, she considers whether the optic nerve resembles the kind of thread that was used to sew on buttons in her home (25/14) and rejects the image of "wiring" to describe the functioning of the brain (62/40). As if to reinforce the point, her reflections on how the doctors remove the tumor from her brother's brain are directly juxtaposed and metaphorically compared with her own plucking of weeds in the garden (40-41/25-26) so as to assimilate the scientific remoteness of brain surgery back into the ordinariness of daily life.

Even though much of the clinical and technical expertise associated with the surgical procedure might lie outside of the lay person's grasp, in other words, the medical risk scenario seems to some extent recuperable into everyday life, at least by way of metaphor. The same is not true, however, of the Chernobyl disaster and the general crisis of modern science and technology it signals. This large-scale, collective and non-local risk scenario, a paradigmatic example of "second-hand non-experience," triggers what the narrator herself recognizes as a crisis of representation, a violent clash between scientific realities and ordinary as well as poetic and religious language. Again and again, the narrator finds herself confronted with the Beckian intrusion of scientific language into everyday thought and the unexpected double meanings and ironies it creates for non-scientific discourse. Chernobyl forces average citizens to acquire an entirely new vocabulary: "So setzen sich die Mütter vors Radio und bemühen sich, die neuen Wörter zu lernen. Becquerel . . . Halbwertszeit, lernen die Mütter heute. Jod 131. Caesium" (35), she reflects at one moment. [So the mothers sit down by the radio and attempt to learn the new words. Becquerel . . . Half-life is what the mothers learn today. Iodine 131. Cesium (27)]. Somewhat later she comments: "Die Physiker fahren fort, in ihrer uns unverständlichen Sprache zu uns zu sprechen. Was sind '15 Millirem fall-out pro Stunde'?" (49). [The physicists continue talking to us in their incomprehensible language. What are 'fifteen millirems fall-out per hour'?" (41)].

But scientific vocabulary not only intrudes upon everyday language, it also reconfigures it in unexpected ways. A word such as "radiation," for example—"Strahlen" or "Bestrahlung" in German—acquires an uncomfortable ambiguity for the narrator as it refers both to the radioactivity that might cause cancer and the medical procedure used to cure it. "Der strahlende Himmel. Das kann man nun auch nicht mehr denken. Auf Bestrahlung können wir aufgrund des histologischen Befunds verzichten, wird der Professor zu dir sagen," she puns to herself at one point (30). [The radiant sky. Now one can't think that anymore, either. We can do without radiation treatment in view of the histological findings, your doctor will tell you (21-22)]. With a quotation from a Brecht poem at another moment, she similarly highlights the clash between poetic references to nature and the environment created by the contemporary risk society: "*O Himmel, strahlender Azur*. Nach welchen Gesetzen, wie schnell breitet sich Radioaktivität aus, günstigenfalls und ungünstigenfalls" (18). [*O heavens' radiant azure*. According to what laws and how quickly does radioactivity spread, at best and at worst?" (9)]. In both cases, lyrical effusions about the beauty of nature take on sinister connotations in the context of nuclear disaster, to the point where the narrator begins to doubt the continued viability of nature poetry: "*Wie herrlich leuchtet mir die Natur*. Vielleicht ist es nicht die dringlichste Frage, was wir mit den Bibliotheken voller Naturgedichte machen. Aber eine Frage ist es schon, habe ich gedacht" (44-45). [*Marvellous Nature Shining on Me!* Perhaps the problem of what to do with the libraries full of nature poems is not the most urgent. But it is a problem all the same, I thought (37)].

Frequent appearances of the word "cloud" [Wolke], which in the context of the novel refers above all to the plume of radioactivity moving westward from Chernobyl, lead the narrator to similar puns, associations and reflections. "Daß wir es 'Wolke' nennen, ist ja nur ein Zeichen unseres Unvermögens, mit den Fortschritten der Wissenschaft sprachlich Schritt zu halten," she comments (36). [Calling it 'cloud' is merely an indication of our inability to keep pace linguistically with the progress of science (27)]. She wistfully remembers her grandmother's time, when a cloud referred to something made up of evaporated water, and responds sarcastically to biblical quotations about clouds on the radio. The use of clouds as a metaphor for whiteness and purity in poetry leads her once again to reflect on the potential obsolescence of the genre:

> Nun aber, habe ich gedacht . . . durfte man gespannt sein, welcher Dichter es als erster wieder wagen würde, eine weiße Wolke zu besingen. Eine unsichtbare Wolke von ganz anderer Substanz hatte es übernommen, unsere Gefühle – ganz andere Gefühle – auf sich zu ziehen. Und sie hat, habe ich wieder mit dieser finsteren Schadenfreude gedacht, die weiße Wolke der Poesie ins Archiv gestoßen. (61)
>
> [But now . . . it should be interesting to see which poet would be the first to dare sing the praises of a white cloud. An invisible cloud of a completely different substance had seized the attention of our feelings – completely different feelings. And, I thought once again with that dark, malicious glee, it has knocked the white cloud of poetry into the archives (55)].[17]

[17] Ute Brandes analyzes Wolf's puns on "Wolke" by arguing that "'Wolke' as an ideal concept is the almost dreamlike symbol of 'die weiße Wolke der Poesie,' derived from Brecht's 'Erinnerung an Marie A.' The cloud here represents the utopian, ethereal realm of pure poetry

More than the discomfort of an ageing writer with the new technological realities that surround her, therefore, the narrator's frequent reflections on the crisis of language that accompanies Chernobyl more broadly signal the demise of a nature aesthetic inherited from Romanticism.

Wolf's insistence on the incompatibility of the total contaminated environment created by the risk society with established patterns of language and metaphor, however, leads to a certain structural paradox in the novel. The idea that an accident such as Chernobyl lies outside the purview of normal modes of linguistic representation comes to a climax toward the end of the narrator's long day, when she watches coverage of the disaster on state-controlled GDR television. Ever more outraged by the prevarications and hypocrisies she sees proffered by scientists and journalists, she is surprised when one of the experts finally admits that the risks of new technologies may not be entirely controllable. In response to a journalist's question as to whether predictions about the functioning of new technologies can ever be error-free,

> nun haben der Moderator und ich zu unserer schmerzlichen Überraschung erleben müssen, daß der sich bei aller Bereitschaft zum Entgegenkommen auf diese Aussage nicht hat festnageln lassen wollen. Nun, haben wir ihn sagen hören. Absolut fehlerfreie Prognosen – die gebe es für einen so jungen Zweig der Technik allerdings nicht. Da müsse man, wie immer bei neuen technischen Entwicklungen, mit einem gewissen Risiko rechnen, bis man auch diese Technik vollkommen beherrschte. (106)
>
> [the moderator and I were forced to learn, to our painful surprise, that this guy – despite his general willingness to be accommodating – was not about to be pinned down to this statement. Well, we heard him say, there was no such thing as an absolutely faultless prognosis in such a young branch of technology. As always with new technological developments, one would have to take certain risks into account until one fully mastered this technology as well. (102-103)]

In reaction to this admission of error and accident as part and parcel of complex technological systems and the new normalcy they create, the narrator insists on the exceptionality of the disaster and the impossibility of ever integrating it into everyday life either linguistically or practically:

> Ich habe ja gewußt, daß sie es wissen. Nur, daß sie es auch aussprechen würden, und sei es dieses eine Mal – das hätte ich nicht erwartet. Mir ist ein Brief-

which floats in a sphere so far removed from this day's reality that it must now be relegated to the archives of sentimentality" (108). See also Saalmann (242-43).

> text durch den Kopf gegangen, in dem ich – beschwörend, wie sonst – irgend jemandem mitteilen sollte, daß das Risiko der Atomtechnik mit fast keinem anderen Risiko vergleichbar sei und daß man bei einem auch nur minimalen Unsicherheitsfaktor auf diese Technik unbedingt verzichten müsse. Mir ist für meinen Brief im Kopf keine reale Adresse eingefallen, also habe ich einige Schimpfwörter ausgestoßen und den Kanal abgeschaltet. (106-107)
>
> [I knew very well that they knew it. Only, I had not expected that they would also say it – be it only this one time. The text for a letter went through my mind in which I—imploringly, how else—was to communicate to someone that the risk of nuclear technology was not comparable to [almost] any other risk and that one absolutely had to renounce this technology if there was even the slightest element of uncertainty. I could not think of a real address for the letter in my mind, so I swore out loud and switched channels. (103)]

While this may seem like a plausible enough stance of resistance to the normalizing functions of official risk discourse, it creates a stark contrast with the narrative of *Störfall* itself, whose very structure is predicated on the comparability and metaphoricity of risks. Not only is the juxtaposition of brain surgery and nuclear accident hardwired into the basic plot configuration, but the narrator's perambulations, conversations and reflections during the day create additional fields of comparison by leading her to (her own and her neighbors' and visitors') memories of the Nazi period and World War II. Reminiscing about the way in which National Socialism endangered populations both inside and outside Germany, she also thinks back to the threats of air raids, typhoid epidemics and displacement that she, her brother and some of her neighbors were exposed (and in some cases succumbed) to. Explicitly, the narrator wonders in her long meditations whether evolutionary tendencies toward domination, aggression and violence anchored deep in the human brain (or at any rate, the male brain, as the more feminist moments of the novel suggest) might in the last instance be responsible for both world wars and large-scale technological disasters. Implicitly, the question that the extended historical memory sequences in the novel raise is whether the pervasively contaminated surroundings of the 1980s really represent unprecedented landscapes of risk, or whether similar total risk environments were already created by world wars earlier in the century.

As if this were not enough, the narrator ends her day after switching off the TV in frustration by turning to fiction; specifically, to Joseph Conrad's high-modernist classic *Heart of Darkness*, which in her reading becomes a more authentic expression of her situation than the commentary on radio and television she has followed all day. The horrors of colonialism in the Belgian Congo, by implication, come to superimpose an additional layer of historical comparison onto the strata of Nazism and World War II. Yet of course they also differ from these latter allusions in that they come to Wolf's narrator in aesthetically mediated form – not as her own memories, but as a novel, and this seems to be the reason that she finds a solace in reading Conrad that she was unable to derive from any other source. Literature, then, is not quite as obsolete in Wolf's risk society as one might have concluded from her narrator's earlier reflections on the questionable relevance of nature poetry. Its function seems to be precisely to provide frameworks of comparison for contemporary risk scenarios whose aesthetic nature allows them to operate in ways that other kinds of comparison cannot. Yet if reading fiction provides a way of reintegrating extraordinary risk into ordinary experience at the end of the novel, the tensions of this moment of narrative closure with the protagonist's climactic assertion that Chernobyl bears (almost) no comparison remain unresolved. In its concluding moments, *Störfall* remains uneasily suspended between a rhetoric that affirms the exceptionality of contemporary technological risk scenarios and one that on the contrary emphasizes their ordinariness from a historical perspective.

Both Carson and Wolf, then, in quite different historical and cultural contexts, rely on comparison as their basic strategy of conveying the new ecological and technological risks of the post-World War II era. Carson explains the dangers of chemical exposure by way of the comparison with the communist scare, nuclear radiation, and biblical apocalypse; Wolf recurs to brain surgery, World War II and colonialism to convey the scope and scale of the Chernobyl disaster. Yet the differences between the two texts are equally striking. Most centrally, the comparative structure remains unambiguous and unproblematic in *Silent Spring*, whereas it is problematized and self-consciously reflected on at every step in *Störfall* and leads to a rhetorical structure whose tensions remain unresolved at the end. It is precisely the historical difference between 1962 and 1987, I would argue, combined with the very different socio-political context of the US and GDR, that accounts for this difference. Carson wants to convey to an unknowing public that it indeed lives in a world of pervasive ecological risks; Wolf, a quarter-century later, acknowledges this by then established idea but also wants to resist its normalization and acceptance, especially in a context of state-controlled scientific and media establishments, which leads her to portray the nuclear explosion as both an unusual accident and the normal day's news. In Wolf, as well as in Beck's theory of the risk society, scientific language therefore assumes a dual role: on one hand, it is the means by which the normality of risk is conveyed; but on the other hand, scientific discourse is also what keeps disrupting the normality of an experience that still tends to rely on culture as a realm separate from science.

3. Irony and the Aesthetic of Uncertainty

Life in a postmodern riskscape of multiple dangers from industry and contamination also lies at the heart of Steven Matheson's experimental short black-and-white film, *Apple Grown in Wind Tunnel: Wind Speed 85 MPH* (2000). "You don't remember this yet. But depending on what you do, some day you might," a male voice says at the beginning, introducing a scenario that seems partly satire and partly near-future science fiction. The film's plot revolves around a woman who broadcasts the procedures for making unusual medical remedies from an unknown location. In a parodic inversion of traditional cooking recipes and homeopathy, these remedies, which provide cures for illnesses ranging from the common cold to cancer, are concocted from the toxic substances spread about the landscape by omnipresent polluting industries, such as "small doses of pesticide [or] effluent from titanium refining processes." Housewives, bikers and retirees begin to collect and exchange these recipes that they hear on ever-changing radio frequencies, and a genuine alternative health network emerges when long-distance truck drivers begin to listen the broadcasts and to communicate the recipes to each other via CB and bulletin boards at roadside cafés. This quiet revolution soon prompts a backlash from the official health industry and government institutions. Some members of the network are arrested and fined for storing hazardous materials, its first official conference is broken up by police for its alleged threat to public health, and pharmaceutical companies appropriate the recipes and attempt to disrupt their free dissemination by restricting knowledge of their ingredients again to an elite of health experts. Yet the network persists in spite of this political oppression, and the film ends with a vision of a strangely utopian society in which large numbers of people begin to recover from their illnesses – by means of cures based on the most toxic substances known to modern chemistry. "This story begins with a mistake. It begins with the presumption of ill health," the voice repeats at two different moments. As it turns out, the "mistake" here is precisely Rachel Carson's central insight that sustained exposure to toxic chemicals in the environment inflicts lasting damage on human health.

At first sight, *Apple Grown in Wind Tunnel* might look like a straight-out satire of Carson's or Beck's idea of totally contaminated environments (when the anonymous broadcaster directs her listeners to toxic harvesting sites, the narrative voice comments, "Soon you realize that maybe the whole landscape is aglow, and that you'd never noticed") with perhaps a more specific barb aimed at the genre of the environmental memoir that has become extremely popular in American literature of the 1990s, from Terry Tempest Williams' *Refuge* to Sandra Steingraber's *Living Downstream* and Susan Antonetta's *Body Toxic*. It might also allude to the cruder satire of Kaufman and Herz' film *The Toxic Avenger* (whose wimpy loser protagonist falls into a vat of toxic chemicals only to be transformed into a hulkingly ugly but righteous hunter after urban villains), Todd Haynes' ambiguously ironic treatment of Multiple Chemical Sensitivity in his film *Safe*, or the thematically broader literary eco-satires of Carl Hiaasen. Yet the precise target of Matheson's irony may be even harder to identify than that of Haynes' *Safe*. Carson's insight that toxins become more concentrated as they accumulate in animal organisms, a process which she illustrates in *Silent Spring* again and again through anecdotes involving the discovery of dead songbirds, is ironically reversed in Matheson's film. "Key substances move through the food chain. They accumulate in high concentrations in some species. Some animals, mainly insects and amphibians, become valued for their medicinal properties: the Los Alamos tree frog, the Hanford grasshopper," the narrative voice explains at one point.

But whether this is intended as a criticism of environmentalist concerns about contaminated ecosystems becomes more and more questionable as the film emphasizes that the remedies, made from waste substances that most of society is eager to rid itself of, can be prepared at home by relatively simple means and obtained without too much effort by those in need. The toxic homeopathy network thereby emerges as an alternative to a health industry that not only keeps its knowledge of cures to itself and its prices for medication high, but that indeed has a vested economic interest in not curing its customers definitively. "[P]harmaceutical companies, HMOs and insurance interests compete globally. Their doctors want your pain. It's a renewable resource. It drives the medical economy by way of an . . . alchemical process: once transformed into symptom, your suffering turns to gold," the narrator points out in one of the moments of most direct social criticism in the film. The unknown woman who first created the toxic remedies, by contrast, "doesn't care about your diagnosis, just about your cure. And she doesn't want to have your pain, or take it away from you. She has no use for it." The emergence of such an alternative health care system, the narrator indicates, leads to a moment in which "the health care equivalent of the crash of 1929 is just around the corner" and consequently to the pharmaceutical industry's appropriation of the "postindustrial cures." Clearly, in scenes such as these, the target of the film's irony is not the environmentalist movement but the pharmaceutical-industrial complex.

Yet *Apple Grown in Wind Tunnel* can by no means be reduced to a satirical attack on the contemporary health care industry – if only because the toxic medicine network itself provides such an ambiguous counterpoint to it. In her last broadcast, the anonymous woman who has started the network indicates that she has done away with her own brain, nerves, chest and stomach by means of her alternative remedies and notes with a chuckle that she has "never felt better," a narrative conclusion that hardly inspires confidence in the therapeutic intent of her brand of medicine. In cinematographic terms, however, this dissolution of the body is no more than a confirmation of the status quo, since the woman in fact never appears on the screen, but is only ever heard as a voice. Neither is any other human body shown, whether in health or in illness, except for hands pointing, writing or handling objects. In some scenes where these hands are shown copying down the recipes, however, another odd dissonance arises: as the female voice spells out the ingredients of each remedy with exact quantities indicated as for a cooking recipe, the hands are shown noting down only the numbers, as if these were some kind of secret code. In addition, the voice punctuates her ingredient lists and directions to toxic sites with what seem to be random sequences of numbers, and toward the beginning and the end of the movie, a hand is shown sliding over reams and reams of pure numerical columns. These scenes make full sense only when one considers the Matheson's multiple allusions to earlier films, which include Orson Welles' *Citizen Kane* and Lizzie Borden's cult classic *Born in Flames*, among others. The particular allusion in the broadcast scenes is to Jean Cocteau's film *Orphée* (1950), in which the famous poet Orphée falls in love with a beautiful and enigmatic aristocrat who turns out to be his own death, and who in the end sacrifices herself for him so as to ensure the poet's artistic immortality. During their acquaintance, Orphée becomes obsessed with strange broadcasts he can only receive on her car stereo, which combine elusive poetic lines with exactly the same kind of random number sequences that the female voice enunciates in *Apple Grown in Wind Tunnel*. The allusion to Cocteau therefore fulfills a dual function. It evokes a story of erotic attraction to one's own death that ironizes the female voice's ostensible commitment to life and health, thereby further undercutting any reading of the toxic network as a serious alternative to corporate health care. But it also, and perhaps more importantly, invokes a well-known narrative model in which the fascination with death is associated with the search for a new type of aesthetic.

The camera's frequent returns to scenes of hands writing numbers by means of the low-tech media of pencil and paper already hints at a combination of scientific abstraction with domestic materiality that is also foregrounded in other scenes. Shots of kitchens in which conventional pots and pans are joined with test tubes and other chemical lab equipment in the production of the new kinds of medication bring science, technological risk and domesticity together in ways unimagined by either Carson or Wolf. But perhaps most importantly, it is Matheson's black-and-white landscape photography that transforms Carson-style risk rhetoric with its emphasis on dead animals, withering plans and blighted landscapes. Matheson's bleak industrial and suburban landscapes stand out by their lyrical beauty: electric poles or highway overpasses are reflected in slightly quavering water puddles, the lights of industrial complexes shimmer in the night, or the camera zooms in on finely veined wings of radioactive insects. These landscapes and details are indisputably beautiful even though they are also contaminated and deadly, indeed, they are attractive in part *because* they are dangerous. Even though the narrative voice of Matheson's film may be ironic, therefore, the photography is not. When the narrator points out that pharmaceutical companies have to appropriate the toxic remedies because "there are no more herbs left to harvest. Where else can [they] turn?," that the new medical knowledge "could make a Superfund site as valuable as a rainforest," and that it has given rise to a new custom of "family outings to abandoned heavy-metal leach fields," it already suggests that the toxic, industrial landscapes of Beck's risk society have in fact become our new nature and our new domesticity. The beauty of the landscape shots and close-ups reinforces this suggestion by stripping it of irony and indicating that this new nature deserves the same visual and verbal celebration as the British Romanticists' Lake District, Thoreau's *Walden Pond* or Ansel Adams' *Yosemite*. Technological and ecological risks and the scientific language by which they are accompanied, therefore, have become an integral part of a new cultural landscape in Matheson's film. Partly ironically and partly seriously, *Apple Grown in Wind Tunnel* portrays a counterculture that turns conventional risk discourse upside down precisely by making it the basis of its definition of the natural, the everyday and the domestic. How literally Matheson wants his viewers to take the workings of this counterculture matters less than the film's broader suggestion that scientific risk discourses have become an inescapable part of culture in the broad sense and that they also offer new possibilities for the aesthetic forms of expression that are usually associated with "culture" in the narrow sense.

Carson and Wolf, as I have shown, both deploy comparison and metaphor to negotiate the relationship between omnipresent scenarios of technological and ecological hazards and an everyday domesticity that both of them portray as threatened by these risks. Carson, in comparing the presence of chemical toxins to nuclear radiation, characterizes it as an exceptional state of crisis that she hopes can be reversed – not through the abandonment of science and technology, but through a different pattern of usage. Wolf, writing a quarter of a century after Carson, is less sure that modern society can move beyond such pervasive riskscapes, and her historical comparisons also suggest that she is more ambivalent than Carson and other environmentalists as to whether humans have not been exposed to similar dangers from non-technological sources in the past. Her own rhetoric of comparison suggests an oscillation between the urgency to portray the risks associated with modern technologies such as nuclear power plants as extraordinary, uncontrollable and in need of elimination, and the at least partial conviction that such encompassing risk environments are (and perhaps have been for a long time) part of human cultures and are not likely to be overcome as part of a new sociocultural normalcy. Matheson's film, made at the turn of the third millennium, seeks to translate this idea of a new toxic normality aesthetically through its uneasy combination of a narrative that lends itself to being read ironically with a photographic style that does not. All three works, I would suggest, can fruitfully be understood as attempts to configure aesthetically what Ulrich Beck has theorized as "second-hand non-experience," the reshaping of everyday life through scientific discourses that include a significant component of uncertainty.

It is in the last instance this uncertainty to which Carson's and Wolf's metaphors as well as Matheson's irony strive to give aesthetic expression. In Carson, the search for an aesthetic of uncertainty is least pronounced as her use of metaphor remains subordinate to her overarching goal of risk communication; but in the reception of *Silent Spring*, it is precisely the most literary passages rather than the technical explanations of pesticide damage that have most successfully imprinted themselves on public awareness. Christa Wolf, approaching broad questions of risk, technology and modernity by means of fiction, is much more directly concerned with finding appropriate aesthetic means of capturing the new uncertainties of everyday experience along with the cognitive, perceptual and linguistic reversals they bring about. As I have shown, *Störfall: Nachrichten eines Tages* ultimately remains suspended between a rhetoric that asserts the uniqueness of the nuclear threat and one that normalizes it by means of a whole range of technological and historical comparisons. Steven Matheson's *Apple Grown in Wind Tunnel: Wind Speed 85 MPH* shifts the inquiry into the aesthetics of risk from metaphoric to ironic narrative structures. Portraying a landscape pervaded by omnipresent toxins, radiation and other forms of pollution, his utopian vision of a new kind of medicine leaves his viewers as unsure of the ultimate target of its irony as they must be in the face of the uncertainties that the risk society surrounds them with on a daily basis. Taken together, the three works outline the cultural development from Carson's alarm call to the perceptual normalization of pervasive ecological and technological hazards in Matheson, at the same time that they foreground the increasingly urgent search for an aesthetic of uncertainty that would be commensurate with contemporary cultures of risk.

Works Cited

Apple Grown in Wind Tunnel: Wind Speed 85 MPH. Dir. Steven Matheson. Video Data Bank, 2000.

Beck, Ulrich. *Der kosmopolitische Blick oder: Krieg ist Frieden*. Frankfurt a.M.: Suhrkamp, 2004.

---. *World Risk Society*. Cambridge, MA: Polity Press, 1999.

---. *Risikogesellschaft: Auf dem Weg in eine andere Moderne*. Frankfurt a.M.: Suhrkamp, 1986.

Beck, Ulrich, Anthony Giddens, and Scott Lash. *Reflexive Modernization: Politics, Tradition and Aesthetics in the Modern Social Order*. Cambridge, MA: Polity Press, 1994.

Born in Flames. Dir. Lizzie Borden. Perf. Honey, Adele Bertei, Jean Satterfield, Florynce Kennedy, and Becky Johnston. First Run Features, 1983.

Brandes, Ute. "Probing the Blind Spot: Utopia and Dystopia in Christa Wolf's *Störfall*." *Selected Papers from the Fourteenth New Hampshire Symposium on the German Democratic Republic*. Ed. Margy Gerber et al. Lanham, MD: University Press of America, 1989. 101-114.

Buell, Frederick. *From Apocalypse to Way of Life: Environmental Crisis in the American Century*. New York, NY: Routledge, 2003.

Buell, Lawrence. *The Future of Environmental Criticism: Environmental Crisis and Literary Imagination*. Oxford: Blackwell, 2005.

---. *Writing for an Endangered World: Literature, Culture and Environment in the U.S. and Beyond*. Cambridge, MA: Harvard University Press, 2001.

---. *The Environmental Imagination: Thoreau, Nature Writing, and the Formation of American Culture*. Cambridge, MA: Harvard University Press, 1995.

Carson, Rachel. *Silent Spring: Fortieth Anniversary Edition*. Boston: Houghton Mifflin, 2002.

Citizen Kane. Dir. Orson Welles. Perf. Orson Welles, Joseph Cotton. Mercury Productions, 1941.

Deitering, Cynthia. "The Postnatural Novel: Toxic Consciousness in Fiction of the 1980s." *The Ecocriticism Reader: Landmarks in Literary Ecology*. Ed. Cheryll Glotfelty and Harold Fromm. Athens, GA: University of Georgia Press, 1996. 196-203.

Douglas, Mary and Aaron Wildavsky. *Risk and Culture: An Essay on the Selection of Technological and Environmental Dangers*. Berkeley, CA: University of California Press, 1982.

Eysel, Karin. "History, Fiction, Gender: The Politics of Narrative Intervention in Christa Wolf's *Störfall*." *Monatshefte* 84 (1992): 284-298.

Finucane, Melissa L., Ali Alhakami, Paul Slovic, and Stephen M. Johnson. "The Affect Heuristic in Judgments of Risks and Benefits." *Journal of Behavioral Decision Making* 13 (2000): 1-17.

Fischhoff, Baruch, Sarah Lichtenstein, Paul Slovic, Stephen L. Derby, and Ralph L. Keeney. *Acceptable Risk*. Cambridge, MA: Cambridge University Press, 1981.

Fischhoff, Baruch, Paul Slovic, and Sarah Lichtenstein. "Lay Foibles and Expert Fables in Judgments about Risks." *Progress in Resource Management and Environmental Planning*. Vol. 3. Ed. Timothy O'Riordan and R. Kerry Turner. Chichester: John Wiley, 1981. 161-202.

Garrard, Greg. *Ecocriticism*. London: Routledge, 2004.

Giddens, Anthony. *Modernity and Self-Identity: Self and Society in the Late Modern Age*. Stanford: Stanford University Press, 1991.

---. *The Consequences of Modernity*. Cambridge: Polity Press, 1990.

Glotfelty, Cheryll. "Cold War, Silent Spring: The Trope of War in Modern Environmentalism." *And No Birds Sing: Rhetorical Analyses of Rachel Carson's Silent Spring*. Ed. Craig Waddell. Carbondale, IL: Southern Illinois University Press, 2000. 157-173.

Hebel, Franz. "Technikentwicklung und Technikfolgen in der Literatur." *Der Deutschunterricht* 41 (1989): 35-45.

Heise, Ursula K. *Sense of Place and Sense of Planet: The Environmental Imagination of the Global*. New York: Oxford University Press, 2008.

---. "Afterglow: Chernobyl and the Everyday." *Nature in Literary and Cultural Studies: Transatlantic Conversations on Ecocriticism.* Ed. Catrin Gersdorf and Sylvia Mayer. Amsterdam: Rodopi, 2006. 177-207.

---. "Science, Technology and Postmodernism." *The Cambridge Companion to Postmodernism.* Ed. Steven Connor. Cambridge, MA: Cambridge University Press, 2004. 136-167.

---. "Science and Ecocriticism." *American Book Review* 18.5 (July-August 1997): 4/6. Repr. in *Mots Pluriels* 11 (1999): http://www.arts.uwa.edu.au/MotsPluriels /MP1199.html and in http://www.asle.umn.edu/archive/intro/heise.html.

Herber, Lewis [Murray Bookchin]. *Our Synthetic Environment*. New York: Knopf, 1962.

Kasperson, Roger E. "The Social Amplification of Risk: Progress in Developing an Integrative Framework." *Social Theories of Risk.* Ed. Sheldon Krimsky and Dominic Golding. Westport, CT: Praeger, 1992. 153-178.

Kasperson, Roger E. et al. "The Social Amplification of Risk: A Conceptual Framework." *Risk Analysis* 8.2 (1988): 177-187.

Kaufmann, Eva. "Unerschrocken ins Herz der Finsternis: Zu Christa Wolfs 'Störfall'." *Christa Wolf: Ein Arbeitsbuch*. Ed. Angela Drescher. Berlin: Aufbau, 1989. 252-269.

Killingsworth, M. Jimmie and Jacqueline S. Palmer. "Millennial Ecology: The Apocalyptic Narrative from *Silent Spring* to *Global Warming*." *Green Culture: Environmental Rhetoric in Contemporary America*. Eds. Carl G. Herndl and Stuart C. Brown. Madison, WI: University of Wisconsin Press, 1996. 21-45.

Latour, Bruno. "Why Has Critique Run out of Steam? From Matters of Fact to Matters of Concern." *Critical Inquiry* 30 (2004): 225-248.

Lear, Linda. "Introduction." *Silent Spring: Fortieth Anniversary Edition*. Boston, MA: Houghton Mifflin, 2002. 10-19.

Löfstedt, Ragnar E. and Lynn Frewer. "Introduction." *The Earthscan Reader in Risk and Modern Society*. London: Earthscan, 1998. 3-27.

Lupton, Deborah. *Risk*. London: Routledge, 1999.

Magenau, Jörg. *Christa Wolf: Eine Biographie*. Berlin: Kindler, 2002.

Orphée. Dir. Jean Cocteau. Perf. Jean Marais, Maria Casares, Marie Déa, and François Peirer. Films du Palais Royal, 1950.

Patrick, Amy M. "Apocalyptic or Precautionary? Revisioning Texts in Environmental Literature." *Coming into Contact: Explorations in Ecocritical Theory and Practice*. Ed. Annie Ingram, Ian Marshall, Daniel J. Philippon, and Adam W. Sweeting. Athens, GA: University of Georgia Press, 2007. 141-153.

Perrow, Charles. *Normal Accidents: Living with High-Risk Technologies*. Princeton, NJ: Princeton University Press, 1999.

Rayner, Steve. "Cultural Theory and Risk Analysis." *Social Theories of Risk*. Ed. Sheldon Krimsky and Dominic Golding. Westport, CT: Praeger, 1992. 83-115.

Renn, Ortwin and Bernd Rohrmann (eds.). *Cross-Cultural Risk Perception: A Survey of Empirical Studies*. Dordrecht: Kluwer, 2000.

Rey, William H. "Blitze im Herzen der Finsternis: Die neue Anthropologie in Christa Wolfs *Störfall*." *The German Quarterly* 62 (1989): 373-383.

Ross, Andrew. *The Chicago Gangster Theory of Life: Nature's Debt to Society*. London: Verso, 1994.

Rudd, Robert L. *Pesticides and the Living Landscape*. Madison, WI: University of Wisconsin Press, 1964.

Saalmann, Dieter. "Elective Affinities: Christa Wolf's *Störfall* and Joseph Conrad's *Heart of Darkness*." *Comparative Literature Studies* 29 (1992): 238-258.

Safe. Dir. Todd Haynes. Perf. Julianne Moore, Peter Friedman, Xander Berkeley. Sony Pictures Classics, 1995.

Sjöberg, Lennart. "Principles of Risk Perception Applied to Gene Technology." *European Molecular Biology Organization* 5 (2004): 47-51.

---. "Are Received Risk Perception Models Alive and Well?" *Risk Analysis* 22 (2002): 665-669.

Slovic, Paul. "Perception of Risk." *Science* 236 (17 April 1987): 280-285. Rpt. in *The Perception of Risk*. London: Earthscan, 2000. 220-231.

---. "Trust, Emotion, Sex, Politics, and Science: Surveying the Risk-Assessment Battlefield." *Risk Analysis* 19 (1999): 689-701. Rpt. in *The Perception of Risk*. London: Earthscan, 2000. 390-412.

Starr, Chauncey. "Social Benefit versus Technological Risk." *Science* 165 (1969): 1232-1238.

The Toxic Avenger. Dir. Lloyd Kaufman and Michael Herz. Perf. Mark Torgl, Mitch Cohen. Troma Entertainment, 1985.

Turney, Jon. *Frankenstein's Footsteps: Science, Genetics and Popular Culture*. New Haven, CT: Yale University Press, 1998.

Weiss, Sydna Stern. "From Hiroshima to Chernobyl: Literary Warnings in the Nuclear Age." *Papers on Language and Literature* 26 (1990): 90-111.

Wolf, Christa. *Störfall: Nachrichten eines Tages. Verblendung: Disput über einen Störfall*. München: Luchterhand, 2001.

---. *Accident: A Day's News*. Trans. Heike Schwarzbauer and Rick Takvorian. New York: Farrar Straus Giroux, 1989.

Wynne, Brian. "May the Sheep Safely Graze? A Reflexive View of the Expert-Lay Knowledge Divide." *Risk, Environment and Modernity: Towards a New Ecology*. Eds. Scott Lash, Bronislaw Szerszynski, and Brian Wynne. London: Sage, 1996. 44-83.

"The habit of saying I:" eigenvalues and resonance

Hanjo Berressem

The essay traces the migrations of the term 'eigenvalue' [*Eigenwert*] and some of its derivatives (such as 'eigenfunction' 'eigenspace,' 'eigenoperation' and 'eigen-organization') between the 'cultures' of mathematics (Hilbert), physics (Schrödinger), systems theory (von Foerster, Luhmann), literary studies (Dilthey), philosophy (Deleuze) and literature (Pynchon), showing how each of them incorporates the term—as well as the terminological cluster that has accrued around it—into its specific operative logic. From the complex field of conceptual resonances that have developed around the term, the essay develops a new understanding of conceptual passages between a multiplicity of cultures understood as resonant fields (Serres, Deleuze).

"the habit of saying I:" eigenvalues and resonances

"The Art of Science: The Two Cultures and Beyond"—the title of the publication of which the following essay is a part—evokes quite programmatically the steady functional differentiation [*Ausdifferenzierung*] of Charles P. Snow's '2 cultures' into a state of what today may be called '2+ cultures.' Rather than taking on a global issue within this field, the following paper will trace one of the countless local histories that compose it; that of the migration of the term 'eigenvalue' [*Eigenwert*] from mathematics to physics, systems theory, literary theory, philosophy and literature. I am not only interested in the 'eigenvalue' of this history, however—although more work needs to be done on such conceptual and terminological migrations within the 2+ cultures debate—but in how it might promote an understanding of conceptual|terminological migrations within the 2+ cultures landscape as movements within a multiplicity of resonant fields.

introduction: resonant fields

Let me begin with three seemingly unrelated stories. The first is about an extremely unscientific experiment that concerns a pair of artificial birds that I keep on my bookshelf and who, when one opens the lid of their little cardboard habitat, chirp as long as their solar batteries hold [image1]. A while ago, a friend played some Steve Reich music from his laptop in the room with the bookshelf while we were discussing the notion of eigenvalues. After about twenty minutes, we heard a chirping from the closed box on the bookshelf that was curiously in tune with the music. We listened for a while, and when we stopped the music, the chirping stopped as well. When we started it again, after about twenty minutes of music, the birds once more started to chirp. When we stopped the music for good, the chirping stopped as well.

The second story comes from Differential Equations and Their Applications by Michael Braun:

> The Tacoma Bridge was built in 1940. From the beginning, the bridge would form small waves like the surface of a body of water . . . . At one point, one edge of the road was 28 feet higher than the other edge. Finally, this bridge crashed into the water below. One explanation for the crash is that the oscillations of the bridge were caused by the frequency of the wind being too close to the natural frequency of the bridge. The natural frequency of the bridge is the eigenvalue of smallest magnitude of a system that models the bridge. This is why eigenvalues are very important to engineers when they analyze structures. (Braun 171-173)

The third story, which is about humans considered as bundles of eigenvalues|eigenfrequencies, comes from the work of the German philosopher Wilhelm Dilthey. The text in question, which deals with the relation between 'a' life and 'its' autobiography, has become so central in studies of the autobiography that it has been extracted from the collected works and published separately as "Das Erleben und die Selbstbiographie." At one point in the text, Dilthey notes that

> [t]he intrinsic values [*Eigenwerte*] experienced in, and only in, the lived experience of the present are accessible to experience in a primordeal way, but they stand juxtaposed to each other without any connection. For each of them arises in the concern of a subject for an object accessible to it in the present . . . . Thus the intrinsic values of the experienced present stand juxtaposed and unconnected; they can only be assessed when they are compared with each other. Anything else described as valuable must be referred back to intrinsic values [*Eigenwerte*]. (*Selected Works* 223)[1]

mathematics: vectors

In his 1904 essay "Grundzüge einer allgemeinen Theorie der linearen Integralgleichungen," the German mathematician David Hilbert introduced the neologism eigenvalue—together with that of 'eigenfunction'[2]—into linear algebra (I have not been able to ascertain whether this is indeed the overall first use of the term. Although there might have been earlier uses, however, it was only after Hilbert that the term became a common coinage not only in the field of mathematics). The context of Hilbert's introduction of the term was the modelization of transformations or operations (in which operators H(x) describe specific programs, such as 'the action of H on x') by way of 1. the identification of those vectors within the transformation|operation that remain invariant and 2. the measurement of the scalar changes of these vectors, such as those brought about by operations of stretching or compression.[3] Hilbert derived the term from the German word 'eigen,' which has a number of meanings, among them 'proper,' 'inherent,' 'intrinsic' 'own' and 'characteristic.' In such transformations|operations, Hilbert called a preserved direction

[1] As the English translation renders eigenvalue either as 'distinctive value' or as 'inherent value,' most of the connotations of the term are lost on the reader.

[2] "Insbesondere in dieser ersten Mitteilung gelange ich zu Formeln, die die Entwicklung einer willkürlichen Funktion nach gewissen ausgezeichneten Funktionen, die ich E i g e n f u n k t i o n e n nenne, liefern" (51); "Wir führen hier noch folgende Bezeichnungen ein: die Nullstellen von $\delta(\lambda)$ mögen die *zum Kern (s,t) gehörigen Eigenwerte* heißen" (64).

[3] Mathematically, an operator is a program [*Rechenvorschrift*] that is performed on a whole function and which transforms this function into another one.

an 'eigenvector' (the eigenvectors of a linear operator are non-zero vectors which, when operated on by the operator, result in a scalar multiple of themselves) and the associated amount by which it has been scaled its eigenvalue (if there is a vector (object x) that has been preserved by H apart from submitting it to a scalar multiplier k, x is called an eigenvector of H with the eigenvalue k). A transformation that does not affect the direction of a vector, Hilbert called an eigenvalue equation [*Eigenwertgleichung*]. The notion of eigenvalue, then, concerns the orientation and the scale of vectors within a mathematical transformation|operation.

The reason why the term could migrate so easily into other fields is that one can generalize from 'vectorial change' to 'systemic change.' As concerns systems—whether these are mathematical, physical and biological—one can say that *eigenvalues|eigenvectors, as they define invariants within systems while these undergo changes, can be used to model systems in relation to changes that they undergo.* Or, if one stresses the processual rather than the systemic side, *as they define invariants within processes, eigenvalues|eigenvectors isolate systems within these processes.* If there is no eigenvector|eigenvalue within a process, the process is not *system*atic or *object*ive in the sense that there is no invariant element that could be said to have undergone the process. In other words, to identify an eigenvector|eigenvalue is to define a system as the cluster of invariant characteristics within a specific process: a system can only be defined *within* a specific process as 'something undergoing change' and it is only within the parameters of the specific process|change that the system can be defined as a system. Any system x, therefore, might be defined as 'the invariance in undergoing the transformational process y.' In this definition, the concept of a structural or ontological essence of a system is replaced by the concept of a system defined as a set of invariants within a process, or, from the systemic side, as a set of 'stable' characteristics during a series of transformations. 'Eigenfaces,' for instance, which are used for computerized face recognition, are sets of statistically standardized face ingredients, or eigenvectors, modeled on the invariants that define specific faces within the multiplicity of processual fluxes that they undergo through time.[4] (As the same technique is also used for handwriting analysis, lip reading, voice recognition and medical imaging, one also talks more generally of 'eigenimages.') Eigenfaces, therefore, are not the ontological essences of

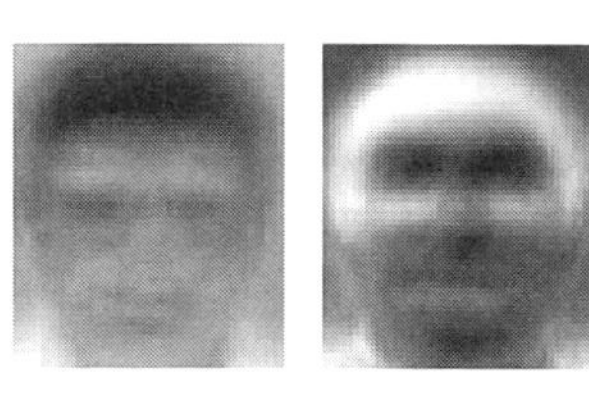

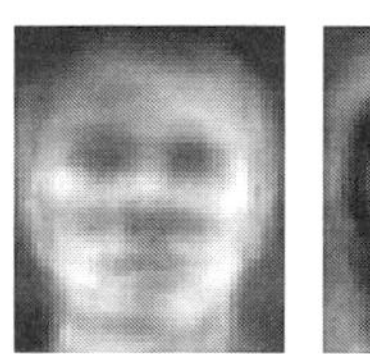

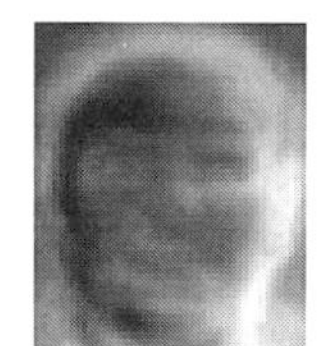

[4] The approach of using eigenfaces for recognition was developed by Matthew Turk and Alex Pentland beginning in 1987. Eigenfaces are derived from the covariance matrix of the probability distribution of the high-dimensional vector space of possible faces of human beings.

these faces. If the sets of invariants that define the original faces were to change, the eigenfaces would change as well. *Mathematically, then, eigenvalues define sets of 'invariants within processes.'*

physics: frequencies and waves

In physics, which is the field to which the term migrated first, eigenvalues are used in a number of contexts that reach from quantum mechanics (the Schrödinger equation from 1926, for instance, is a wave equation in which the eigenvalues are the energy levels of the system in Ψi)[5] to questions of how to model 'systems|objects

[5] "To obtain specific values for energy, you operate on the wavefunction with the quantum mechanical operator associated with energy, which is called the Hamiltonian. The operation of the Hamiltonian on the wavefunction is the Schrodinger equation. Solutions exist for the time-independent Schrodinger equation only for certain values of energy, and these values are called 'eigenvalues' of energy. ... While the energy eigenvalues may be discrete for small values of energy, they usually become continuous at high enough energies because the system can no longer exist as a bound state. For a more realistic harmonic oscillator potential (perhaps representing a diatomic molecule), the energy eigenvalues get closer and closer together as it approaches the dissociation energy. The energy levels after dissociation can take the continuous values associated with free particles" (http://hyperphysics.phy-astr.gsu.edu/hbase, / quantum/scheq.html#c5). "The wave defined by the Schrödinger equation does not only transport energy, but also particles, such as an electron. Like all other waves, it has two components, which can not be measured individually, however, [which means that] the two waves' components do not represent a measureable field. The measureable particle is always in both conponents of the wave. It is also not important, whether at a certain position one or the other component is stronger" (www.quantenwelt.de/quantenmechanik/wellenfunktion/schrodinger.html; my translation). "Der Hamiltonoperator ist ein zentrales mathematisches Objekt der Quantenmechanik. Er beschreibt die dynamischen Eigenschaften eines Systems und erscheint z. B. in der Schrödingergleichung. Man kann den Hamiltonoperator semi-heuristisch aus der Hamiltonfunktion der klassischen Mechanik ableiten, indem man die dynamischen Variablen durch die entsprechenden quantenmechanischen Operatoren ersetzt. Der Hamiltonoperator ist benannt nach William Rowan Hamilton. Die Eigenvektoren sind die stationären Zustände des Systems, die zugehörigen Eigenwerte E die entsprechenden Energien . . . . Der Spektralsatz garantiert, dass die Energien reell sind und die linear unabhängigen Eigenvektoren eine Basis des Hilbertraums bilden. Je nach System kann das Energiespektrum diskret oder kontinuierlich sein. In der Tat weisen Systeme häufig neben einem diskreten Energiespektrum auch ein energetisch höherliegendes Kontinuum auf. Ein Beispiel dafür ist ein endlicher Potentialtopf, in dem gebundene Zustände mit diskreten negative Energien und freie Zustände mit kontinuierlich verteilten, positiven Energien auftreten" (http://de.wikipedia.org/wiki/Hamiltonoperator). The Hamiltonian function, then, describes "a dynamical system (as a pendulum or a particle in motion) in terms of generalized coordinates and momenta. It is equal to the total energy of the system when time is not explicitly part of the function" (http://against-the-day.pynchonwiki.com/wiki/index.php?title=ATD_149-170).

in process' in terms of their invariant internal frequencies, or 'eigenfrequencies.'[6] In fact, Hilbert's inspiration for the term eigenvalue might well have come from Hermann von Helmholtz' 1863 study *Die Lehre von den Tonempfindungen als Physiologische Grundlage für die Theorie der Musik*, in which von Helmholtz coined the word '*Eigentöne*' ['proper tones,' such as the "eigentone of the resonator" [*Eigenton des Resonators*] (75)] to designate "tones of highest resonance" [*Töne stärkster Resonanz*] (150).[7]

One of the practical uses of the computation of eigenfrequencies concerns the testing of materials. Metal objects, for instance, resonate on a specific eigenfrequency when they are brought to vibrate. This is especially important for tuning forks, which need to resonate precisely at the 'concert pitch' of 440 Hertz.[8] More generally, if the frequency of an industrially produced object is different from that of the other objects within the same series, this points to a material irregularity within the object.

As they are nothing but temporal and local stabilities within a larger processual field, internal frequencies invariably resonate with and respond to the landscape of frequencies in which they are suspended. In my story about the crickets, for instance, it seemed that the frequency of the vibrations that were triggered by the music were 'in tune' with the eigenfrequency of the crickets, which seemed to have picked up and responded to these vibrations, as when a radio or a tv picks up signals|waves transmitted on a 'tuned-into' frequency within the specific medium they travel through. As Friedrich Cramer notes in *Symphonie des Lebendigen: Versuch einer allgemeinen Resonanztheorie*, all resonance phenomena rely on such material media: "[e]ine Welle bedarf eines Mediums, um sich fortzupflanzen, für die Wasserwelle ist es das Wasser, für den Schall im allgemeinen die Luft, für das Licht das elektromagnetische Feld (früher sagte man Äther), für Erdbeben die Erdkruste" (53). In acoustic registers, "Schallwellen werden durch das Medium Luft übertragen, das zum Mitschwingen gebracht wird und seinerseits diese Schwingungen auf ein schwingungsfähiges Gebilde geeigneter Frequenz übertragen kann" (65). In or-

6 "Eine Eigenfrequenz eines schwingfähigen Systems ist eine Frequenz, mit der das System nach einmaliger Anregung schwingen kann. Bei Vernachlässigung der Dämpfung fallen die Eigenfrequenzen mit den Resonanzfrequenzen des Systems zusammen" (http://www.ldw.de/index2.php?area=1&np=56,0,0,0,0,0,0,0).

7 See also: "Jedes einzelne einfache Wellensystem pendelartiger Schwingungen existiert als ein für sich bestehendes mechanisches Ganzes, verbreitet sich, setzt andere elastische Körper von entsprechendem Eigenton in Bewegung, ganz unabhängig von den gleichzeitig sich ausbreitenden anderen einfachen Tönen von anderer Tonhöhe, die aus derselben oder einer anderen Tonquelle hervorgehen mögen" (81-82).

8 On the tuning fork, see Helmholtz 68.

der to resonate with the eigenfrequency of the crickets, the music had to be quite literally 'on the same wavelength' as that of the habitual chirping of the crickets.[9]

The Tacoma event is one of the most famous examples of frequency interference, and a striking image of the fact that "*Resonanz überträgt Energie*" (54). During the event, the interference pattern—as Cramer notes, "Resonanz besteht immer darin, daß Schwingungen miteinander in Wechselwirkung treten und sich überlagern" (53)—resulted in a resonance that was, from a systemic point of view, destructive for the bridge: the eigenfrequency of the wind triggered a positive feedback loop within the eigenfrequency of the bridge. These dynamics were caused by the fact that when "Wellen gleicher Frequenz ineinander laufen, so verstärken sich ihre Schwingungsbäuche" (53). This amplification led to a point at which the internal stresses of the 'system of the bridge' became so strong that its material|structural threshold of tolerance was passed, which caused it to give way, losing its structural stability and its definition as an invariant system.[10] The Tacoma event, then, was one of the cases in which "Schwingungen innerhalb eines Systems sich so sehr synchronisieren, so stark in Gleichtakt geraten, daß ihre Amplituden sich gegenseitig hochschaukeln und so stark werden, daß sie das ganze System zum Zerplatzen bringen" (80). Like all catastrophes, such a "Resonanz-Katastrophe kommt immer unerwartet, plötzlich, unvorhersagbar" (84). In general terms, what happened was that the synchronization of the eigenfrequency of the wind with the eigenfrequency of the bridge brought about a change in the overall landscape of frequencies. For the bridge it meant that it lost its alignment—what Humberto Maturana and Francisco Varela would call its 'structural coupling'—with the landscape of frequencies that it was suspended in and resonated with. Complex systems—the largest of which is the 'system of nature' in general—are full of such catastrophes, most of which are invisible to the human eye. Each cancellation of a turbulence within a body of water—as an invariant system within the overall multiplicity of fluxes|currents—by a current that re-distributes the overall landscape of turbulences, for instance, is such a catastrophic event. In his novel *Cryptonomicon*, Neal Stephenson describes the break-up of such a system from the position of that system; in this case a 'stable' current of air:

> From the wind's frame of reference, it [the wind] is stationary and the hills and valleys are moving things that crumple the horizon and then rush towards it and

[9] In order to react to minute perturbations|disturbances within the medium, the receiving system needs to be in a sensitized, 'far-from-equilibrium' state.

[10] Such feedback loops, in which the values increase exponentially, are also called positive feedback loops. "Die Energie wird bei periodischer Anregung optimal übertragen und im System gespeichert. Durch die Speicherung und weitere Energiezufuhr schwingt das System immer stärker, bis die Belastungsgrenze überschritten ist" (http://de.wikipedia.org/wiki / Resonanzkatastrophe).

> then interfere withit and go away, leaving the wind to sort out the consequences later on down the line . . . . If there was more stuff in the way, like expansive cities filled with buildings, or forests filled with leaves and branches, then that would be the end of the story; the wind would become completely deranged and cease to exist as a unitary thing, and all of the aerodynamic action would be at the incomprehensible scale of micro-vortices around pine needles and car antennas. (622)

Not all frequency interferences, however, need to be catastrophic, and Stephenson also describes a number of harmonious instances of frequency interferences, such as the harmonies made up of the wind when it meets a staircase, which becomes, in the process, a musical scale.

> Wind and water have been whipped into an essentially random froth by the storm. A microphone held up in the air would register only white noise – a complete absense of information. But when that noise strikes the long tube of the staircase, it drives a physical resonance that manifests itself in Waterhouse's brain as a low hum. The physics of the tube extract a coherent pattern from meaningless noise . . . . Waterhouse experiments by singing the harmonics of this low fundamental tone . . . . Each one resonates in the staircase to a greater or lesser degree. It is the same series of notes made by a braß instrument. By hopping from one note to another, Waterhouse is able to play some passable bugle calls on the staircase. (230)

Finally, high-energy resonance dynamics can also be used to serve specific purposes. In *Spiderman 3,* when Spiderman surrounds Venom with a ring of metal rods, which he then strikes, causing them to resonate with Venom's eigenfrequency, this brings about Venom's very own resonance catastrophe. (The susceptibility of aliens to such resonance catastrophes is a common science-fiction motif, exploited, amongst others, in the movie *Mars Attacks*).

In the trajectory from von Helmholtz to Schrödinger and further to Cramer, eigenvalues are related to the fields of wave-, frequency- and resonance phenomena. While *in both the mathematical and the physical context, a system's eigenvectors|eigenvalues|eigenfrequencies denote its invariants within a specific dynamics* (*its coherence*), the physical field adds a decidedly 'material' aspect to the term.

literary studies: philosophy of life

While the migration of the term eigenvalue from mathematics to physics is quite straightforward, its migration to Dilthey's 'philosophy of life' [*Lebensphilosophie*] is more convoluted and enigmatic. In fact, it is not quite clear to what degree Dilthey relied on Hilbert's mathematical use of the term, although his many other scientific references and his repeated use of mathematical models make a direct connection at least plausible, especially since his theory of the autobiography is ultimately a theory of how humans perceive themselves as what Gregory Bateson would call "*pattern*[*s*] *through time*" (13). Still, there are other conceiveable backgrounds for Dilthey's use of the term, such as Immanuel Kant's *Groundwork of the Metaphysics of Morals* [*Grundlegungen für eine Metaphysik der Sitten*]; a text that, in its differentiation between beings who have "ends in themselves" and beings who can be used as mere "means" (41),[11] has become the main source for the common current meaning of eigenvalue, which is 'intrinsic value' as opposed to 'use value.' What makes Kant a less plausible background for Dilthey, however, is that Kant does not use the term eigenvalue directly—in this context, today's use of the term is a back-projection onto Kant's text—and the fact that Dilthey's use of eigenvalue goes well beyond the differentiation between 'intrinsic value' and 'use value.'[12] Another possible source for Dilthey is *The Ego and his Own* [*Der Einzige und sein Eigentum*] from 1844, in which the German philosopher Max Stirner promotes any number of 'eigen' compounds—the fact that the German language abounds with possibilities to create 'eigen' compounds certainly facilitated Stirner's somewhat inflationary use of the term—in the context of the description of his anarcho-individualistic philosophy, such as *Eigentum* [property], *Eigenheit* [individual characteristic: 'ownness'] and *der Eigene* [the individual: 'the owner'].[13] Somewhat curiously, however, almost the only 'eigen'term Stirner does *not* use is eigenvalue. Let me assume for a moment, then, that Dilthey knew of Hilbert's use of the term and that the origin of his use of the term lies indeed with Hilbert. Why would this mathematical reference be useful for Dilthey?

In his essay, Dilthey operates with two oppositional models of time that are roughly equivalent to Henri Bergson's differentiation between *temps* and *duree*. The formal|mathematical model of time is that of an infinitely dense|fast succession of temporal moments|instants (*temps*). In what might be an indirect reference to the mathematical procedure of the 'Dedekind cut,' Dilthey illustrates this model by the infinite subdivision of a continuous line by cuts, each present|instant being conceptualized as a mathematical point 'without extension.' "If we think of time in abstraction from what fills it, then its parts are equivalent to each other," Dilthey notes. The "content of the lived experience" (*Selected Works* 93), as something that

[11] The categorical imperative is to "treat himself and all others *never merely as means* but always *at the same time as ends in themselves*" (41).

[12] Another background for the development of this meaning is the work of Jakob von Uexküll. See especially Karl Friederichs' proposal to replace Uexküll's term 'Umwelt' by 'Eigenwelt' in his book *Ökologie als Wissenschaft von der Natur oder biologische Raumforschung*.

"constantly changes" (93), however, is not 'contained' in the chronological succession of these 'formal' cuts, because it relies on a continuous duration [*duree*] that takes place between the cuts that mark the infinitely dense succession of instantaneous presents ['*ist*'].

Unlike the mathematical present, the perceived present has an extension, which is why systems with a perceptual apparatus have no access to instantaneous presents (except, of course, in the case of humans as an abstract|mathematical notion). In perceived time, even the smallest subdivided moment *is*, or *has* duration. As Dilthey notes, "in this continuity, even the smallest part is linear; it is a sequence that elapses. There is never an *is* in the smallest part" (93). If one were to consider events as the temporal medium of perception and define an event as the smallest loosely coupled element of [ap]perceived time, the event would not lie in the infinitely small, arrested instant—or 'matter-of-fact'—of a pure present, but in the "filling of a moment of time with reality" (93) and thus in the opening up of a pure present onto a, however short, duration. In other words, a durational present lies always between a past and a future.

In opposition to abstract, mathematical time, Dilthey calls this perceived|perceptual time "[c]oncrete" or "real time" (94). Its temporal modality—as the medium of perceptual, memnonic and cognitive processes|formations—is not that of infinitesimal instants|fractions of 'being' but of processual 'becoming.' In Dilthey's words, perceived time is always a "course" (94) whose spatial modality is linear|continuous and thus it is inherently historical rather than punctual|instantaneous. Or, in the words of Cramer, "[d]ie Welt existiert nicht, sie ereignet sich" (22).

This internal split in|of time creates the paradox that although "we always live in the present" (Dilthey, *Selected Works* 94), since "the present never is, and even the smallest part of the continual advance in time contains the present and a memory of what was just present, *what* is present as such is never experienceable by us" (94). Invariably, memory both 'comprehends' and 'stretches' the single, instantaneous moment: "The present never *is*; what we experience as present always contains the memory of what has just been present" (216). Having split time into a physical and a perceptual form that operate simultaneously, although on completely separated plateaus, Dilthey turns to the difference between the perceiving system's physical and psychic organization; that system being, for Dilthey, invariably the human being. For Dilthey, the human being's physical|energetic coherence—its organic "course of . . . life" (93)—is processual and *continuous*: "the existence of a person during his lifetime. The property of uninterrupted constancy belongs to that existence" (93). As systems with a perceptual apparatus, humans can never experience this intensive, physical flux: "[w]e experience both the changes in what-just-was and that such changes are occurring, but we do not experience the flux itself" (217). The subject's material 'course of life' consists of the accumulation|accretion of single, lived moments|actions into a series of moments. In this systemic history, the system's life is a continuous process to which an overall, invariant eigenvector|eigenvalue might be attributed as concerns its 'systematicity': the succession|accumulation of lived instants makes up the systemic coherence that the human being quite literally *is*, as well as the value that defines the collective operations that the system performs|undergoes in the envelope of its operational space|time. As Jean-Pierre Dupuy notes,

> [a] given network usually possesses a multiplicity of self-behaviors, or as they are sometimes called 'attractors' (a term borrowed from dynamical systems theory), and converges toward one or another depending on the initial conditions of the network. The 'life' of a network can thus be conceived as a trajectory through a 'landscape' of attractors, passing from one to another as a result of perturbations or shocks from the external world. (553) [14]

Physical continuity and an overall physical eigenvalue, however, do not necessarily imply the notion of what Dilthey calls an "ideal unity of the parts of life" (94), which is related to the system's *psychic* rather than to its *physical* coherence. This ideal unity, in which single events—as the smallest modules of processual time in which one can identify specific psychic eigenvalues—are aligned and subsumed|integrated into an increasingly complex set, is that of the overall "meaning" (94) of a specific human life. This meaning is constructed by perceptual syntheses, which integrate series of single events (rather than single 'matters-of-fact' which would pertain to the system's systemic course of life) with their specific eigenvalues into larger sequences of individual memories, and thus align the single eigenvalues within|into an overall psychic continuity. As Dilthey notes, "[E]ach life has its own sense. It consists in a meaning-context in which every remembered present possesses an intrinsic value [*Eigenwert*], and yet, through the nexus of memory, it is also related to the sense of the whole" (221). Although it is its 'attribute,' the psychic continuity operates on a different level than the physical continuity of the system's course of life and is therefore independent of it. As Dilthey notes, "*independently* thereof [the course of life], there is an experienceable connection;" a sense of a whole that is "quite independently of their succession in time" (93, my emphasis). Although this connection is the attribute of its physical continuity, it operates in a completely different register, because it is the system's overall integrated, psychic eigenvalue. The meaning of a life, then, is the architecture of psychic events that is abstracted from the system's concrete physical|organic course of

[14] See also: "[t]he dynamics of the network may therefore be said to *tend toward* an attractor, although the latter is only a product of these dynamics" (554).

life. This psychic eigenvalue is the basis for the creation of what the German sociologist Niklas Luhmann calls a "möglichkeitsorientierte Eigenwelt" (684).

In Dilthey, the construction of meaning—as a reflected, psychic coherence—is related to the agency of the 'observer,' who comes to function, in the case of the autobiography, predominantly as the 'self-observer.' In fact, the autobiography is the literary medium in which the subject [→ autopoietic system] contemplates this correlation from a point of retrospective overview, in a coupling of the retrospection [*Nachträglichkeit*] of meaning with the retrospection of writing.[15] As Dilthey notes, "[a]utobiography is merely the literary expression of the self-reflection of human beings on their life-course" (222). Symptomatically, as Philippe Lejeune notes in *Le Pacte autobiographique*, the guarantee of the operation of the genre lies in a contract between the author and the reader that is based on the use of the author's proper name, or 'Eigenname,' in the text, which Gottlob Frege defines as "[d]ie Bedeutung eines Eigennamens ist der Gegenstand selbst, den wir damit bezeichnen; die Vorstellung, welche wir dabei haben, ist ganz subjektiv; dazwischen liegt der Sinn, der zwar nicht mehr subjektiv wie die Vorstellung, aber doch auch nicht der Gegenstand selbst ist"(42).

It holds for both physical and psychic levels that, if their continuity remains unbroken through the set of changes that the system undergoes, they come to define the system as a coherent|continuous aggregate. In the case of the physical level, this pertains to the coherence[s] of the organism. In the case of the psychic level, it pertains to the invariance of the integrated overall eigenvalue. It bears stressing again that *although these are radically separated realms, the set of psychic eigenvalues (psychic reality) is invariably the 'attribute' of the physical coherence[s] (lived reality), behind which one cannot reach.* As its editor, Bernhard Groethuysen, notes about the manuscript of Dilthey's essay,

> Das Manuskript über das Erleben lag in einem Umschlag . . . . Oben auf dem Umschlag befindet sich die folgende Bemerkung: "Leben ist ein Teil des Lebens überhaupt. Dieses aber ist, was im Erleben und Verstehen gegeben ist. Leben in diesem Sinne erstreckt sich sonach auf den ganzen Umfang des objektiven Geistes, sofern er durch das Erleben zugänglich ist. Leben ist nun die Grundtatsache, die den Ausgangspunkt der Philosophie bilden muss. Es ist das von innen Bekannte; es ist dasjenige, hinter welches nicht zurückgegangen werden kann. Leben kann nicht vor den Richterstuhl der Vernunft gebracht werden." (359)

[15] Dilthey's philosophy of life 'evokes,' in many ways, Humberto Maturana and Francisco Varela's theory of autopoiesis.

If life forms the basis [*Grundtatsache*] of thought, the meaning of a life lies in the relation of the organic|physical course of life [→ what Deleuze will call the realm of 'organic, passive syntheses'] to an observed coherence [→ what Deleuze will call 'perceptual, active syntheses']. The writer of an autobiography narrates|translates his|her own course of life into a meaningful assemblage. The medium of this assembly are a multiplicity of singular, unrelated micro-events—small processes with their own eigenvalues—that stand next to each other without relation. From this multiplicity, the writer creates a coherent, relational continuum to which an overall value|meaning (this time a 'hermeneutical' rather than a scientific|mathematical one) can be attributed. Single, lived events are first subsumed into memories—this is an automatic process, to which everyone who has tried to willingly forget a specific event or chain of events can attest—and then consciously integrated *into* and organized *by* a retrospective narrative that creates the present overall value of the 'system of his|her life:' "a nexus that is determined by the relation of the significant moments of life to my present interpretation of it" (Dilthey, *Selected Works* 95). This narrative continuity makes it possible to bestow meaning upon the subject's life *as* an organically coherent aggregate that reflects upon itself. This meaning is assembled on a level of 'active syntheses' [→ contemplation] that lies *beyond* the processes of organic syntheses [→ bio-physical eigen-organizations] and the succession of single|lived instants, as well as *beyond* the eigenvalues of single events. To become meaningful, the perceived moments need to be subsumed according to an algorithm that contains|expresses the coherence of the overall 'system of events.' Out of many smaller processes with their specific eigenvalues, this algorithm creates the system's overall psychic eigenvalue; its psychic course of life:

> Since, with the advance of time, that which has been experienced constantly grows and receeds ever more, memories of our own life-course are formed . . . . In all these memories, some state of being is linked with its milieu of external states of affairs, events, and persons. An individual's life-experience results from the generalization of what has thus accumulated. It arises through procedures that are equivalent to induction. The number of instances on which the induction is based constantly increases in the course of a lifetime. (154)

Without this integration, the value of each event would shift according to its momentuous relation to other moments, which means that from this "point of view of value" (223), each life would consist of nothing but a multiplicity of singular—unorganized|unvectorized—events which form

> a chaos of harmonies and dissonances. Each of them is like a chord that fills a present, but they have no musical relation to each other . . . . Only the category

> of meaning overcomes the mere juxtaposition or subordination of the parts of life to each other. (223)

Comparable to the way in which chaotically|irregularly distributed photons are entrained into a highly 'vectorized' laser beam, the autobiographical construction of the 'meaning of a life' entrains events into a vectorized and vectorizing narrative. Symptomatically, Cramer notes a similar entrainment within sets of molecules or neurons. The operation of the brain, for instance, is based on the fact that "die Neuronen ihre Impulse *synchronisieren*" (134); a process Cramer calls "erzwungene Resonanz" (166). Every organism is 'forced' in this way because "[e]in Organismus ist aus unzähligen Biorhythmen mit bestimmten Eigenfrequenzen zusammengesetzt, die, damit der Organismus funktioniert . . . in Resonanz stehen müssen" (174). In fact, "[a]lle *normalen, stabilen Systeme* sind Überlagerungen von Schwingungen, die in Resonanz stehen" (56).[16] Within this entrained|composed arrangement, "[d]ie Elemente, die Unterstrukturen, die Subsysteme sind ihrerseits schwingende Systeme mit Eigenzeiten" (80) and "Eigenrhythmus" (108).

The integration of singular moments into an overall coherence organizes|reduces the multiplicity of unrelated psychic values into an overall harmonic set of psychic eigenvalues; into the habit of thinking of oneself as a 'self.'

systems theory: eigenorganizations

When the term eigenvalue migrates to 'systems theory'—as a set that includes 2nd order cybernetics and radical constructivism as subsets—it is used to deal precisely with the field that Dilthey had left untheorized as the 'organic continuity' of the subject's course of life. For systems theory, the question is not only about the observer's psychic organization, but also about its physical|organic organization|stabilization, which is, as Dilthey had noted, 'independent' of its perceptual stabilization|synthesis. The project of system's theory is to understand the observer not only as an observing, but also as a living system.

For this project, systems theory integrates the notion of eigen*values* into a theory that treats physical systems as eigen*organizations*, and their (both physical *and* psychic) characteristics as their specific 'eigenbehaviour' ['*Eigenverhalten*,' or '*Eigenschaften*']. As in the mathematical and the physical contexts, systems theory does not consider eigenorganizations and their specific 'eigenspaces' as given, es-

[16] See also William H. Calvin. *The Cerebral Cod: Thinking a Thought in the Mosaics of the Mind*, who talks of the processes of inter-neuronal "entrainment" (33) and "recruiting" (35).

sential objects, but as the result of recursive operations|iterations that have settled into 'stable' attractors.[17] As Heinz von Foerster notes,

> Pragmatisch entsprechen diese [operationellen] Zustände der Errechnung von Invarianten, seien es Objekt-Konstanz, perzeptive Universalien, kognitive Invarianten, Identifikationen, Benennungen etc. Natürlich müssen hier auch die klassischen Fälle von Ultrastabilität und Homoeostase erwähnt werden. (*Wissen und Gewissen* 75)

In all of these contexts, eigenvalues denote "funktionale," "operationale" und "strukturelle" "*Gleichgewichtszustände*" (107).[18]

In a mapping of concepts from non-linear dynamics onto the theory of eigenorganizations, von Foerster visualizes the genesis of an autopoietic system as the contraction|condensation of a field of operational density around a specific point of 'strange attraction.' "About 20 years ago," von Foerster remembers,

> there was an explosion of renewed interest in . . . recursive operations, as one discovered that many functions develop not only stable *values* but also a stable *dynamic*. One called these stabilities 'attractors,' apparently a leftover from a teleological way of thinking. Since one can let some systems march through the most diverse Eigen behaviors by making simple changes in the parameters, one soon stumbled onto a most interesting behavior that is launched by certain parametric values: the system rolls through a sequence of values without ever repeating one, and even if one believes one has taken one of these values as the

[17] Eigenspaces are the respective fields in which the transformations|operations take place and can be described, the "eigenspace of a given transformation being the set of all eigenvectors of that transformation that have the same eigenvalue, together with the zero vector, which has no direction" (http://en.wikipedia.org/wiki/Eigenvalue). On these recursive operations see especially Heinz von Foerster. *Wissen und Gewissen: Versuch einer Brücke* 260.

[18] Luhmann draws directly on von Foerster when he notes in *Die Wissenschaft der Gesellschaft* that "Eigenwerte" (113)—and thus "stabile Eigenzustände" (95)—are created through recursive operations of 'the observation of observations,' in which the repetition of the operation of identification succeeds in "condensing" ["*kondensieren*"] (312) "das für identisch Gehaltene" (312). Through these condensations, "errechnet das System seine 'Eigenwerte' und identifiziert Identität als Zeichen für solche Eigenwerte, und über Eigenwerte kann es dann Eigenverhalten organisieren" (312). In this way, "[d]as System festigt 'Eigenzustände' und gewinnt damit eine dynamische Stabilität" (320). This 'morphogenetic stability' does not only define autopoietic systems: as every system is dependent on the dynamics of its praxis, the system of science, for instance, "kann . . . keine Eigenrationalität entwickeln" (670). All it can produce are "[r]elativ stabile 'Eigenzustände'" (670). As Luhmann notes, "[s]tatt auf letzte Einheiten zu rekurrieren, beobachtet man Beobachtungen, beschreibt man Beschreibungen. Auf der Ebene zweiter Ordnung kommt es erneut zu rekursiven Vernetzungen und zum Suchen von 'Eigenwerten,' die sich in den weiteren Operationen des Systems nicht mehr verändern" (717).

> initial value, the sequence of values cannot be reproduced: the system is chaotic. (*Understanding* 316)

Symptomatically, von Foerster notes specifically that the notion of chaotic, strange attractors is a new modelization of eigenorganizations (*Wissen und Gewissen* 261).

Systemically, eigenvalues come into being when certain functional|formal operations reproduce the same value upon every re-entry into the functional system or 'formalism.' As von Foerster notes, "under certain conditions there exist indeed solutions which, when reentered into the formalism, produce again the same solution. These are called 'eigen-values'" (Segal 164). Because systems model their current behavior on their former behavior, they are computationally recursive:

> one is no longer interested [so much] in . . .[the system's] computational capacities, only in its 'self-behaviors' (or *eigenbehaviors*, to use the hybrid English-German phrase of quantum mechanics, from which systems theory borrowed the concept) . . . . Like every automaton with an internal state, a network calculates its state in the next period as a function of its state during the current period . . . . Note that these external events come to acquire meaning in the context of the network as a result of the network's own activity: the *content*—the meaning—which the network attributes to them is precisely the self-behavior, or attractor, that results from them. Obviously, then, this content is purely endogenous, and not the reflection of some external 'transcendent' objectivity. (Dupuy 553)

Such recursive operations|processes inform the processes of the creation and the gradual stabilization of a system that is actually nothing but the *product* of these recursive operations; nothing but a set of invariant values operating within a specific computational envelope. These purely repetitive|iterative operations|processes, whose differences remain within a restricted numerical|formal envelope, can be considered as 'computational habits,' which means that the creation of a computational attractor can be understood as a case of 'habit-formation;' a topic to which I will return in the section on 'eigenphilosophy.' In fact, systemic habits—sets of stable eigenvalues or 'eigencharacteristics'—are the only indicators of systematicity and the only ground from which to make predictions about a specific system's eigenbehavior:

> This result, that there are Eigenvalues, is the only thing we can rely on. For then an opaque machine begins to behave in a predictable way, for as soon as it has run into an Eigen state, I can of course tell you, for example, if this Eigen state is a period, what the next value in the period is. Through this recursive closure and only through this recursive closure do stabilities arise . . . . What is

> fascinating is that while one can observe these stabilities it is in principle impossible to find out what generates these stabilities. One cannot analytically determine how this system operates, although we see that it does operate in a way that permits us to make predictions. (*Understanding* 316-317)

In analogy to the definition of eigenvalues as systemic invariants within processes, processes of physical eigenorganization define systemic invariants. Through iterated|habitual bio-chemical and informational operations, a system slowly contracts around an 'eigenattractor,' separating itself from the infinitely complex multiplicity of the processual space that surrounds it. Through this generative process, it takes on, for other eigenorganizations, the character|contour of a stable object with specific properties that are set against this very background. As Luhmann notes, "objects appear as repeated indications, which, rather than having a specific opposite, are demarcated against 'everything else'" (*Art as a Social System* 46). Biologically, then, eigenorganizations might be described as 'the things formerly known as living beings' or also as 'living beings under radically constructivist conditions.' They are self-organizing—'self' being a German word that is closely related to the term 'eigen'—temporally equilibrious systems that operate within a material medium from which they have separated themselves through recursive bio-chemical and informational operations. As von Foerster notes, "Eigenvalues, because of their self-defining (or self-generating) structure imply topological closure ('circularity')" (*Understanding* 265). Accordingly, Luhmann can call a chapter in his study *Observations of Modernity* "Kontingenz als Eigenwert der modernen Gesellschaft." (As with the original Dilthey texts, these connotations are lost in translation, which gives the chapter as "Contingency as Modern Societies' Defining Attribute.")

When an eigenorganization encounters habitual|recurrent operations in its medium|surroundings, it isolates these habitual operations from the set of random stimuli|perturbations that are constantly provided by that medium, calling the result of these cut-outs stable 'objects:' As von Foerster notes, "what is referred to as 'objects'. . .in an observer-excluded (linear, open) epistemology, appears in an observer-included (circular, closed) epistemology as 'tokens for stable behaviors'" (*Understanding* 261) or as "tokens for Eigen-functions" (261). In Dilthey's words, "Materialität ist nichts anderes als Widerstandsfähigkeit durch die tastende Hand. Materie ist nichts Reales, sie ist bloss Geschöpf des Tastsinn" (Lessing and Rodi 53-54).[19] In perceptual terms, things [objects] and other eigenorganizations [sub-

[19] See also: "When we ascribe an objective value to something, this means merely that various values can be experienced in relation to it. When we ascribe an instrumental value [*Wirkungswert*] to an object, we merely designate it as capable of bringing about some value at a subsequent point of time. All these are purely logical relations into which a value, experienced in the present, can enter" (223).

jects] that eigenorganizations compute from the irritations provided by the medium—a process of differentiation also called 'observation'—are once more not 'essential, given objects,' but the result of the perceptual system's own internal informational procedures and operations in response to sets of incoming invariant irritations|stimuli. From these sets, eigenorganizations extrapolate, through their own 'eigenbehavioral' operations, specific 'habits|eigencharacteristics' (such as, for instance, the seemingly objective notions of 'color,' 'hardness|elasticity' or 'movement') that define these 'other' objects|subjects for them. This extrapolation of the object|other through the system's eigenbehavior entails what von Foerster calls the conversion of an infinite causal nexus into a finite one:

> Eigenbehavior generates discrete, identifiable entities. Producing discreteness out of infinite variety has incredibly important consequences. It permits us to begin naming things. Language is the possibility of carving out of an infinite number of possible experiences those experiences which allow stable interactions of yourself with yourself. (quoted in Segal 143-144)

For the perceiving system, therefore, subjects|objects are only secondarily the result of the actions|behavior of these subjects|objects. Because they impinge upon the perceiving system only as sets of invariant stimuli, they are first of all the result of the perceiving system's own eigenbehavior, which is why they are merely "*tokens* for eigen-behaviours . . . . In the cognitive realm, objects are the token names we give to our eigen-behaviour. This is the constructivist's insight into what takes place when we talk about our experience with objects" (von Foerster in Segal 143). Or, as Luhmann puts it, "[d]ie wahrgenommene Welt ist mithin nichts anderes als die Gesamtheit der 'Eigenwerte' neurophysiologischer Operationen" (*Die Kunst der Gesellschaft* 15). As recursive operations need a specific duration, eigenvalues are

> recursive distinctions that unfold—and can only unfold—over time, even as they can only be experienced in the nano-moment of the present . . . . In fact, Luhmann suggests that we might follow Mead and Whitehead, who "assigned a function to identifiable and recognizable objects, whose primary purpose is to bind time. This function is needed because the reality of experience and actions consists in mere event sequences, that is, in an ongoing self-dissolution." (Wolfe 46)

Ultimately, objects|subjects are "nothing but the *eigenbehaviors* of observing systems that result from using and reusing their previous distinctions" (Luhman "Deconstruction as Second-Order Observing" 768). As von Foerster notes,

> Eigenvalues have been found ontologically to be discrete, stable, separable and composable, while ontogenetically to arise as equilibria that determine themselves through circular processes. Ontologically, Eigenvalues and objects, and likewise, ontogenetically, stable behavior and the manifestation of a subject's 'grasp' of an object cannot be distinguished. (*Understanding* 266)

autopoiesis: cognition

The conceptual relations between 2nd order cybernetics, radical constructivism, systems theory, and the theory of autopoiesis form an extremely dense, heterarchically organized set. In fact, "Eigenorganisationen" (von Foster *Understanding,* 296) as "seltsame Attraktoren" (296) are quite literally what Maturana|Varela call 'autopoietic systems;' living unit(ie)s that are informationally|operationally separated from their medium, but energetically open to it.

> Autopoietische Systeme sind organisatorisch geschlossen und energetisch offen, ich würde sie 'Eigenorganisationen' nennen. Natürlich hätten extraterrestrische Beobachter, die die frühen Stadien der Lebensentstehung auf dieser Erde beobachteten, sie in Ermangelung eines besseren Ausdrucks als 'fremdartige Attraktoren' bezeichnet. (*Wissen und Gewissen* 296)

Manuel De Landa draws on these ideas when he notes that autopoietic systems are "dynamical systems that endogenously generate their own *stable states* (called 'attractors' or eigenstates'), and they grow and evolve *by drift*" (De Landa 63). One definition of autopoietic systems, therefore, is that *autopoietic systems are eigenorganizations that they result from recursive, habitual operations that are part of an evolutionary landscape|medium in which they do not stop creating|maintaining a border between themselves and the processual surroundings in which they operate.*

In shorthand, the theory of autopoiesis might be described as systems theory with a biologial and cognitive spin. As in radical constructivism, 2nd order cybernetics and systems theory, what the autopoietic system experiences as reality is constructed through internal processes|operations of sensation|perception and cognition, the latter of which Maturana|Varela do not use in the classically philosophical sense as the field of disembodied thought processes, but as the more general field of all of those processes that allow a system to respond to and interact with its medium and with other systems in its specific ecological niche.

The arguably most important notion within the theory of autopoiesis is the fundamental split it introduces into the 'system of life.' *All autopoietic systems are operationally|informationally closed.* They are defined by the "*operational closure* in their organization" and by the fact that "their identity is specified by a network of dynamic processes whose effects do not leave that organization" (Maturana and Varela 89). (In this context, *organization* denotes a system's abstract relational blueprint, while *structure* denotes the physical embodiment of that abstract pattern). In informational|operational terms, autopoietic systems are self- or eigenorganized and therefore self-contained entities that are fundamentally separated from their outside. This informational|operational closure, however, is not enough to qualify autopoietic systems as living systems. Even while they are informationally|operationally closed off from the world, they are *energetically open to the world*, if this world is understood not as the one they *construct* (and, in the case of humans, the one they *project* as the mental image [*Vorstellung*] of the world) but as the one in which and with which they are *constructed* and to which they are corporeally immanent; the world in which their living unfolds and is realized. *In energetic terms, autopoietic systems are dissipative, temporally neg-entropic systems that interact with a multiplicity of intensities and whose biological history is subjected to a general entropic slope.* Autopoietic systems, then, are simultaneously *closed* and *open*. As von Foerster notes "[d]er Begriff der Selbstorganisation ist vielleicht der allgemeinste Begriff für die Beschreibung dieser faszinierenden Prozesse, die in organisatorisch *geschlossenen*, energetisch (thermodynamisch) aber *offenen* Systemen auftreten" (*Wissen und Gewissen* 296, my emphasis).

As 'eigenorganizations,' autopoietic systems function as differentiators between the informationally closed field of operation|information and the energetically open field of irritation|intensity. They emerge from an energetic multiplicity through processes of contraction and stabilization; through the repeated, habitual operation of drawing operational|informational distinctions (recursive computation) and of the re-insertion of these distinctions into the system. As the result|totality of these recursive operations, autopoietic systems operate in their eigenspace with the aim of maintaining an overall, operationally integrated eigenorganization and of remaining within the range of operations that allows them to remain systems. Although processes of eigenorganization define the systems in the first place (the systems are strictly the result of their internal operations), their operation is dependent on energetic stimuli, and thus on changes in the medium they operate|live in. If these become too strong to react to them operationally|informationally, the systems will, like the Tacoma Bridge, no longer operate within the tolerances that define their range of 'structural change|morphogenesis' and they lose their overall coherence. In other words, when their eigenvalues|eigenorganizations are challenged, systems are in danger of losing their structural stability. The results of this loss are either physical death|dissolution or psychic dissolution|schizophrenia.

Within both systems theory and the theory of autopoiesis, the operational field of eigenorganizational processes extends from the physical to the psychic. Even the logic of language is implicated in an 'eigenlogic' because communication relies on an evolutionary co-adaptation within which two subjects have aligned and habituated their relations. In terms of co-evolution, language forms the ground|medium on which, as von Foerster notes, "zwei Subjekte, die miteinander rekursiv interagieren . . . stabile Eigenverhaltensweisen ausbilden, die in der Sicht eines Beobachters als Kommunikabilien (Zeichen, Symbole, Wörter usw.) erscheinen müssen" (*Wisse und Gewissenn* 279-280). Within these dynamics, the linguistic code|language forms the attractor around which communicative actions are organized. "Die Sprache kann . . . als ein emergentes Verhaltensmuster, also ein Eigenverhalten, als ein fremdartiger Attraktor oder besser als ein konstruktiver Transaktor verstanden werden, der zwei Autonomien zu einer verschmilzt" (298). As von Foerster notes in an article dedicated to Niklas Luhmann,

> [t]he word 'behavior,' as well as 'conduct,' 'action,' etc., does imply the recognizability of regularities, of 'invariants' in the temporal course of the action . . . . These invariants, these 'Eigen behaviors' arise through the recursively reciprocal effect of the participants in such an established social domain . . . . **Communication is the Eigen behavior of a recursively operating system that is doubly closed onto itself.** The essential thing about the topology of a double closure is that . . . through the hetarchical [sic!] organization that comes with it, the fascinating possibility exists of allowing operators to become operands and operands to become operators. (*Understanding* 321)

philosophy habits

One reason for the turn of philosophy towards Darwinism in the 19th century is that the concept of adaptation as habit-formation provides a model of how to think the eigenorganization and the maintenance of 'patterns through time' without recourse to an essentialist logic. In the wake of this 'biological turn,' 'habitual philosophies,' such as many versions of materialism, process philosophy and pragmatism, conceptualize the human subject as an arrangement of structural invariants defined within an operative envelope of habitual processes. This processual subject, which has all of the characteristics of an eigenorganized, autopoietic system, is staked against the 'essentialist subject' promoted by more idealist philosophies.

Darwinism, however, is only one part of a more general shift within the hard sciences that takes place in the 19th century; the shift from thinking in terms of structure to thinking in terms of process and habit-formation. In his essay "Design and Chance," for instance, Charles S. Peirce defines natural laws as 'nothing but' natural habits; a sentiment echoed by William James in "Psychology: A Briefer Course," in which he uses processes of habit-formation to describe neurological activity.[20] As Brian Massumi notes, "[b]oth Hume, the inventor of empiricism, and C. S. Peirce, the inventor of the pragmatism further developed by James, argued that nature does not follow laws. *Laws follow nature.* What nature does is generate surprises and contract habits" (247). Moving into the scientific world of the late 20th century, Massumi asks "[c]an't the self-organizations of matter described by chaos theory also be considered habits? Aren't they inhumanly contracted habits of matter? . . . Is there a difference in kind or only a difference in mode or degree between the inhuman habits of matter and the human ones?" (237). In fact, in differenting between natural and cultural habits, Massumi describes almost literally the 'habitual computation' that von Foerster had defined as the operational logic of eigenorganization. In Massumi's words, "[h]abit is an acquired automatic self-regulation. It resides in the flesh. Some say in matter. As acquired, it can be said to be 'cultural.' As automatic and material, it can pass for 'natural'" (11).

One of the most recent philosophies that conceptualize the subject as a habitualized process is that of the French philosophers Gilles Deleuze and Félix Guattari. Although they do not mention the terms eigenvalue and eigenorganization directly, they repeatedly reference the theory of autopoiesis and they let the genesis of structures (ontology) revolve around processes of 'emergence' and habit-formation. Although it is mostly Guattari who provides the directly scientific references, many of Deleuze's descriptions of the coming-into-being of the subject resonate directly with the notion of eigenorganization, as when he stresses that "[t]o speak of the subject now is to speak of duration, custom, habit, and anticipation . . . Habit is the constitutive root of the subject" (Deleuze *Empiricism* 92). When Deleuze stresses that nature also operates by habit-formation, he does so from within a project that aims at reversing, like Dilthey, the classical relation between life and thought|reason. "Life will no longer be made to appear before the categories of thought; thought will be thrown into the categories of life," he notes programmatically (*Cinema* 189). Following David Hume and echoing Peirce, Deleuze argues that although habit-formation is a strictly probabilistic, inductive operation—"habit as a probability" (*Empiricism* 66)—it is, at the same time, a principle; the *principle of probability*: "[t]he paradox of habit is that it is formed by degrees and also that it is a principle of human nature: 'habit is nothing but one of the principles of nature,

[20] See for instance: "May not the laws of physics be habits gradually acquired by systems." Charles Sanders Peirce, "Design and Chance" 553; see also: "[t]he laws of Nature are nothing but the immutable habits which the different elementary sorts of matter follow in their actions and reactions upon each other. In the organic world, however, the habits are more variable than this." William James. *Psychology: The Briefer Course* 137. Also, "[n]ature exhibits only *changes*, which habitually coincide with one another so that their habits are discernible in simple 'laws.'" William James, *The Works of William James* 74

and derives all its force from that origin'" (66).[21] As Peirce had noted, "[i]nduction infers a rule. Now, a belief of a rule is a habit. That a habit is a rule active in us, is evident . . . every belief is of the nature of a habit . . . . Induction, therefore, is the logical formula which expresses the physiological process of formation of a habit" (Peirce *Elements of Logic*, 387).

Deleuze stakes habit-formation against the 'principle' of the 'law of reason' much in the same way that the hard sciences stake the principle of habit-formation against the principle of immutable natural laws. If reason is merely the effect of the principle of habitformation, it can no longer be a principle: "The principle is the habit of contracting habits . . . . Habit is the root of reason, and indeed the principle from which reason stems as an effect" (*Empiricism* 66), or, "as Bergson said, habits are not themselves natural, but what is natural is the habit to take up habits" (44). In fact, as the *result* of natural|habitual operations, reason is what has commonly been understood as its 'other,' "an affection of the mind. In this sense, reason will be called instinct, habit, or nature" (30).

Deleuze ties his philosophical reversal to a theory of the emergence of the subject that relies on both physical (primary) and psychic (secondary) habit-formation as the operation according to which "the subject is constituted within the given" (*Empiricism* 104). In this conceptualization, Deleuze distinguishes, like Dilthey, von Foerster and Maturana|Varela, between a physical, energetic or also intensive level and an informational, operational and integrational level, which he relates to what he calls perceptual|active and organic|passive syntheses:

> perceptual syntheses refer back to organic syntheses . . . . We are made of contracted water, earth, light and air – not merely prior to the recognition or representation of these, but prior to their being sensed. Every organism, in its receptive and perceptual elements, but also in its viscera, is a sum of contractions, of retentions and expectations . . . by combining with the perceptual syntheses built upon them, these organic syntheses are redeployed in the active syntheses of a psycho-organic memory and intelligence. (*Difference* 73)

Organically, subjects are formed through systemic contractions [→ condensations] within an elemental multiplicity that Deleuze calls 'plane of immanence' or also 'plane of construction.' These processes of contraction constitute "the primary habits that we are; the thousands of passive syntheses of which we are originally com-

[21] As the result of iterations, habit forms one of the three aspects of the 'trinity of thought:' "We may say, therefore, that hypothesis produces the *sensuous* element of thought, and induction the *habitual* element. As for deduction . . . this may be considered as the logical formula for paying attention, which is the *volitional* element of thought." Charles Sanders Peirce, "Book III: Critical Logic" 387. In fact, mental facts consist of "conceptions, desires . . . expectations, and habits." Charles Sander Peirce, "Book III: Unpublished Papers" 333.

posed" (74). By modeling organic eigenorganization as the contraction of habits—"What organism is not made of elements and cases of repetition, of contemplated and contracted water, nitrogen, carbon, chlorides and sulphates, thereby intertwining all the habits of which it is composed?" (75)—Deleuze turns habit-formation into the philosophical version of what von Foerster had described as the eigenorganization of systems through recursive, 'habitual' computations. As James had noted, because "the law of habit . . . is a material law" (150), "[t]he philosophy of habit is thus, in the first instance, a chapter in physics rather than in physiology or psychology" (*Psychology* 138).[22]

If systemic|organic eigenorganizations are indeed contractions of physical habits— "[h]abit draws something new from repetition, namely difference . . . . In essence, habit is contraction" (Deleuze *Difference* 73)—one must, in order to define a system, find the invariant vectors within a process that allow one to talk of something or someone "contracting a habit" (74). Passive syntheses, which constitute our "habit of living" (74), share with Dilthey's idea of a physical 'continuity' that they "constitute . . . the *living present*" (81, emphasis added). Even during the 'passive' processes of contraction, however, the habitual system as an 'emergent self'[23] feedbacks with the forces of contraction through what Deleuze calls sub-individual processes of 'contemplation' so that it is "simultaneously through contraction that we are habits, but through contemplation that we contract" (74).[24] Even operations that seem to be 'natural' are, in actual fact, of the class of processes Peirce calls "inattentive habit[s]" (Peirce *Survey* 328) and they operate on the same unconscious level as 'perceptual judgements:' "As soon as we find that a belief shows symptoms of being instinctive, although it may seem to be indubitable, we must suspect that experiment would show that it is not really so" (Peirce *Issue* 297). While both conscious and unconscious reasoning are cases of 'habit-formation,' "[t]he term 'reasoning' ought to be confined to . . . fixation of one belief by another as is reasonable, deliberate, self-controlled" (293), although there are also unconscious reasonings: "there are, besides perceptual judgements, original (i.e. indubi-

[22] See also: "[t]he soul is a material body, the body is a thing, the subject is just an object, physiology and psychology is just physics. And, consequently, the senses are reliable." Michel Serres, *The Birth of Physics* 49.

[23] In reference to the 'emergent self' see also Deleuze and Guattari's references to Daniel Stern's *The Interpersonal World of the Infant: A View from Psychoanalysis and Developmental Psychology*. Further contextualizations of the term run from Stuart Kauffman to Maturana and Varela.

[24] "The passive synthesis of habit constituted time as the *contraction* of instants under the condition of the present, the active synthesis of memory constitutes it as the *interlocking* (*emboîtement*) of the presents themselves" (*Difference* 81, revised translation). On passive and active syntheses, as well as on habit-formation in Deleuze, see also Jay Lampert, *Deleuze and Guattari's Philosophy of History*.

table because uncriticized) beliefs of a general and recurrent kind, as well as indubitable acritical inferences" (296).

Despite the fact that Peirce differentiates between the fields of habitualization and what he calls 'natural dispositions'—by which he presumably means a set of 'pre-habitual' biological parameters: "*habit-change*; meaning by a habit-change a modification of a person's tendencies toward action . . . : It excludes natural dispositions, as the term 'habit' does, when it is accurately used . . . the effects of habit-change last until time or some more definite cause produces new habit-changes" (*Survey* 327)—, even natural dispositions in the sense of natural organizations might well be considered as the result of organic micro-habits. As Peirce himself notes, "habit is by no means exclusively a mental fact. Empirically, we find that some plants take habits. The stream of water that wears a bed for itself is forming a habit" (342).

The sub-individual processes of contemplation are similar to Dilthey's psychic eigenvalues, with the important difference that for Deleuze, the 'psychic' realm extends much more deeply into sub-individual levels and that it 'shades'—like Peirce's realm of conscious reasonings—into unconscious, bio-chemical 'contemplations,' which are responsible for the constitution of a "passive self"—which Deleuze also calls a "hecceity" (Deleuze and Parnet 132) or an assemblage (98)—on what Deleuze calls a "plane of immanence or consistence" (94). Ultimately, this passive self is nothing but the 'body of resonance' of specific habits. It "contemplates and contracts the individuating factors of such fields [the 'pre-existing fields of individuation'] and constitutes itself at the points of resonance of their series" (*Difference* 276).

Deleuze's description of the passive self and of the passive syntheses that bring it about resonates directly with the delineation of eigenorganizations as invariants within bio-chemical processes, or, in Deleuze's terminology, 'modifications':

> These thousands of habits of which we are composed—these contractions, contemplations, pretensions, presumptions, satisfactions, fatigues; these variable presents—thus form the basic domain of passive syntheses. The passive self is not defined simply by receptivity—that is, by means of the capacity to experience sensations—but by virtue of the contractile contemplation which constitutes the organism itself before it constitutes the sensations. This self, therefore, is by no means simple: it is not enough to relativise or pluralise the self, all the while retaining for it a simple attenuated form. Selves are larval subjects; the world of passive syntheses constitutes the system of the self, under conditions yet to be determined, but it is the system of a dissolved self. There is a self wherever a furtive contemplation has been established, whenever a contracting machine capable of drawing a difference from repetition functions somewhere.

> The self does not undergo modifications, *it is itself a modification*. (78-79, emphasis added)

Active, individual(izing) syntheses emerge from the field of the passive syntheses and from "our thousands of component habits" (75). Unlike passive syntheses, which seem to be intensive but consist 'in actual fact' of pre-individual, unconscious|inattentive and productive micro-integrations, active syntheses are related to "the principle of representation" (81) and to conscious processes of integration, which means that they operate on a conscious informational|operational level that has its own agenda: "Whereas active synthesis points beyond passive synthesis towards global integrations and the supposition of identical totalisable objects, passive synthesis, as it develops, points beyond itself towards the contemplation of partial objects which remain non-totalisable" (101).

While Deleuze and Dilthey are systemically compatible, therefore, they differ in that Deleuze is not only concerned with the constitution of meaning from the 'ideal' continuity of a system's physical course of life, but also with how to reach back from within representational, active syntheses, such as those of memory, to the seemingly intensive, passive syntheses that are, in actual fact, inattentive biochemical integrations. In relation to time|memory, the question is how, given that passive syntheses are sub-representative, to reach back to them 'despite' the representational operations of memory. In other words, "whether or not we can penetrate into [*pénétrer dans*] the passive synthesis of memory; whether we can in some sense live the being in itself of the past in the same way that we live the passive synthesis of habit" (84, modified English translation).[25]

In a semiotic register, this translates into the question of whether one can penetrate into the level of what Deleuze calls 'natural signs,' which are

> founded upon passive syntheses; they are signs of the present, referring to the present in which they signify. Artificial signs, by contrast, are those which refer to the past or the future as distinct dimensions of the present . . . . Artificial signs imply active syntheses – that is to say, the passage from spontaneous imagination to the active faculties of reflective representation, memory and intelligence. (77)[26]

[25] The original French runs: "Si nous pouvons pénétrer dans la synthèse de la mémoire." It is thus not a question of penetrating the passive syntheses, as implied by the English translation, but of penetrating into the passive syntheses from the level of active syntheses.

[26] In Peirce's semiotic terms, "[t]he deliberately formed, self-analyzing habit . . . is the living definition, the veritable and final logical interpretant. Consequently, the most perfect account of a concept that words can convey will consist in a description of the habit which that concept is

Artificial signs emerge from natural signs in the same way that attentive|conscious integrations ('molar integrations') emerge from inattentive|unconscious ones ('molecular integrations'): "below the level of active syntheses, [there is] the domain of passive syntheses which constitutes us, the domain of modifications, tropisms and little peculiarities" (79). As with Dilthey, "[b]eneath the general operation of laws . . . there always remains the play of singularities" (25). Beneath the psychic reality of meaning lies "the lived reality of a sub-representative domain" (69). Or, in more positive terms, the reality of meaning emerges from the reality of life similar to the way in which reason emerges from habit-formation. Accordingly, "we must regard habit as the foundation from which all other psychic phenomena derive" (78). As Deleuze notes in the preface to the English translation of *Empiricism and Subjectivity,*

> [w]e start with atomic parts, but these atomic parts have transitions, passages, 'tendencies,' which circulate from one to another. These tendencies give rise to *habits*. Isn't this the answer to the question 'what are we?' We are habits, nothing but habits – the habit of saying 'I.' Perhaps there is no more striking answer to the problem of the Self. (x)

As both passive and active syntheses rely on habit-formation as eigenorganization, the subject is "a habitus, a habit, nothing but the habit in a field of immanence, the habit of saying I" (Deleuze and Guattari 48).

Even while, from an ontogenetic perspective, habit-formation promises an evolutionary freedom and the possibility of 'unconditioned' change even within a set of systemic constraints, in the 'practice of the subject' habits can easily harden into stable, quasi-essential routines. This is especially true for its cultural practice and for its habits of thought. If habits become addictive—too much of a habit, that is—the system loses the plasticity it had won as a result of the replacement of essence by habit. As Bateson notes,

> in the ongoing life of the organism there is a process of sorting, which in some of its forms is called 'habit formation.' In this process, certain items, which have been learned at 'soft' levels, gradually become 'hard' . . . . The converse of 'habit formation' . . . is a form of learning which is always likely to be difficult and painful and which, when it fails, may be pathogenic. (Bateson 138)[27]

calculated to produce" (*Survey* 342). In fact, "there remains only habit, as the essence of the logical interpretant" (*Survey* 334).

[27] See also William James' *Psychology: The Briefer Course*.

Spanned out between ontological freedom and constraint, as well as between structural stability and morphogenesis, habits function somewhat like norms. Even if human beings are, as James notes, nothing but "walking bundles of habits" (*Psychology* 145) and thus free to change, habits are also "the enormous fly-wheel of society, its most precious conservative agent. It alone is what keeps us all within the bounds of ordinance" (145). As Peirce noted, inductive habit-formation define both the real and the imaginary realm: "*reiterations in the inner world—fancied reiterations—if well-intensified by direct effort, produce habits,* just as do reiterations in the outer world; *and these habits will have power to influence actual behaviour in the outer world*" (*Survey* 334). These fancied reiterations, or, as Peirce also calls them, "[t]he formation of habits under imaginary action" (*Issues* 294), allow for evolutionary plasticity in the relation of the system to its irritational medium:

> consciousness may be defined as that congeries of non-relative predicates, varying greatly in quality and in intensity, which are symptomatic of the interaction of the outer world—the world of those causes that are exceedingly compulsive upon the modes of consciousness, with general disturbances sometimes amounting to shock, and are acted upon only slightly, and only in a special kind of effort, muscular effort—and on the inner world, apparently derived from the outer, and amenable to direct effort of various kinds with feeble reactions; the interaction of these two worlds chiefly consisting of a direct action of the outer world upon the inner and an indirect action of the inner world upon the outer through the operation of habits. (*Survey* 342-343)

As a belief is a habit—"a deliberate, or self-controlled, habit is precisely a belief" (330)—the oscillation between belief and doubt is the one between habitualization and de-habitualization: "Belief is not a momentary mode of consciousness; it is a habit of mind essentially enduring for some time, and mostly (at least) unconscious; and like other habits, it is (until it meets with some other surprise that begins its dissolution) perfectly self-satisfied. Doubt is an altogether contrary genus. It is not a habit, but the privation of a habit" (Peirce *Pragmatism* 279) Even while habits are less strict than immutable laws, therefore, too often subjects cannot 'kick the habits' that they *consist of.*

literature: the spine of reality

Although it would be a small step from philosophy to literature, when the term eigenvalue migrates to the work of American writer Thomas Pynchon, it comes directly from mathematics. While it is almost too much of a cliché to bring up the work of Thomas Pynchon in the context of "The Art of Science: The Two Cultures and Beyond"—why not Neal Stephenson's *Baroque Cycle* or William Gibson's *Idoru*?—my Pynchon reference is somewhat inevitable, not only because one of the characters in Pynchon's novel *V.* is the dentist-philosopher Dudley Eigenvalue (this could easily be a throwaway reference in the work of Pynchon), but also because Pynchon uses the name to set up a larger mathematical conceit.

Dilthey had shown that for humans, only the successful correlation of the subject's physical course of life with psychic eigenvalues can ensure its internal coherence as a system radically split into physics and 'metaphysics.' The correlation of physical and psychic reality relies on the ability to integrate 'lived' life into a horizon of 'meaningful' life. In *V.*, Eigenvalue laments precisely his inability to perform this integration. Confronted with the seemingly contingent vicissitudes of 'historical' events, Eigenvalue feels that he has lost an overall historical sense:

> perhaps history this century, thought Eigenvalue, is rippled with gathers in its fabric such that if we are situated, as Stencil seemed to be, at the bottom of a fold, it's impossible to determine warp, woof, or pattern anywhere else. By virtue, however, of existing in one gather it is assumed there are others, compartmented off into *sinuous cycles* each of which come to assume greater importance than the weave itself and destroy any continuity. Thus it is that we are charmed by funny looking automobiles of the '30's, the curious fashions of the '20's, the peculiar moral *habits* of our grandparents . . . . We are accordingly lost to any sense of continuous tradition. Perhaps if we lived on a crest, things would be different. We could at least see. (Pynchon *V.* 155-156, emphasis added)

The quote aligns 'eigenvalues,' 'sine waves' and 'habits' in a complex topological image. On the crumpled surface of history—a surface that is defined by a variable curvature—eigenvalue is caught in a historical valley. Lacking an overview and thus a continuous sense of history, all he can make out are seemingly unrelated, 'dissonant' events without, as Dilthey would say, musical harmony. Curious fashion-moments and glimpses of people with weird habits spin around in 'sinuous cycles,' forming a chaotic multiplicity that is not a good starting point for either historiography or autobiography. In fact, this multiplicity marks what has been theorized as the 'end of history' and as the loss of the coherence of psychic reality.[28]

[28] "Für die Analyse kompliziert sich überlagernder Schwingungen gibt es ein bestimmtes mathematisches Verfahren, das ganze, scheinbar unregelmässige 'Überlagerungsgebirge' von Wellenstrukturen in die einzelnen Sinusschwingungen aufzulösen gestattet: die Fourier-Analyse. Das Verfahren ist immer dann angezeigt, wenn eine Vielkomponenten-Struktur mit tausenden von Eigenfrequenzen verstanden werden soll . . . . Im Prinzip lassen sich. . .alle komplexen Schwingungsstrukturen in die Eigenfrequenzen auflösen, aus denen sie zusam-

In *Against the Day*, Pynchon's novel about anarchists and 'state operators' as well as an extended meditation on the nature of waves—whether electrical, optical or historical—, Pynchon returns to the concept of eigenvalues. In fact, he brings the discussion of eigenvalues back to the moment of their conception, in a scene that has the fictional character Yashmeen Harcourt discuss "[t]he nontrivial zero of the ζ-function" (Pynchon *Against* 604) with David Hilbert in Göttingen. "Might they be correlated with eigenvalues of some Hermitian operator yet to be determined?" (604), Yashmeen asks Hilbert. In the ensuing conversation, she interrupts a sentence begun by Hilbert, providing a term that relates mathematical eigenvalues directly both to the physical invariants that make up 'processual people' and the psychic invariants that make up what these people construct as their reality, before Hilbert concludes the sentence by going back to talk pure mathematics:

> "Apart from eigenvalues, by their nature, being zeroes of *some* equation," he prompted gently.
>
> "There is also this . . . spine of reality." Afterward she would remember she actually said "*Rückgrat von Wirklichkeit.*" "Though the members of a Hermitian may be complex, the eigenvalues are real. The entries on the main diagonal are real. The ζ-function zeroes which lie along Real part = 1/2, are symmetrical about the real axis, and so . . . ." She hesitated. She had *seen it*, for the moment, so clearly. (604)

As the invariant vector that turns a process into a 'process undergone by x,' the 'spine of reality' holds eigenorganizations together both physically and psychically. As eigenorganizations, humans have a physical, energetic spine as well as a psychic, informational|operational (and, in the case of humans: observational) spine that is radically different from the physical one—in the sense that relations are external to their terms—although, as embodied, it remains 'attributed' to the physical coherence of the system and emerges from it.

mengesetzt sind" (Cramer 57). This complexity is precisely what, in *The Crying of Lot 49*, Mucho Maas undoes with his 'spectrum analysis.' If a synthesizer assembles single frequencies into complex architectures—"Put together all the right overtones at the right power levels so it'd come out like a violin" (142)—Mucho can do the same in reverse. Listen to anything and take it apart again. Spectrum analysis, in my head, I can break down chords, and timbres, and words too into all the basic frequencies and harmonics, with all their different loudnesses, and listen to them, each pure tone, but all at once" (142). In fact, Mucho can align all of the single iterations of a string of words—considered as assemblages of frequencies—across time: "Everybody who says the same words is the same person if the spectra are the same only they happen differently in time, you dig? But the time is arbitrary. You pick your zero point anywhere you want, that way you can shuffle each person's time line sideways till they all coincide. Then you'd have this big, God, maybe a couple of hundred million chorus. . .and it would all be the same voice" (142).

It is another instance of what Pynchon calls, in *The Crying of Lot 49*, the "high magic to low puns"(129) when the novel also mentions Hilbert's 'Spectral Theory,' an inclusive term for theories extending the eigenvector and eigenvalue theory of a single square matrix, which "requires a vector space of *infinite* dimensions. His co-adjutor Minkowski thinks that dimensions will eventually all just fade away into a *Kontinuum* of space and time" (*Against* 324). The pun lies in the fact that this 'spectral theory' is analogous to Pynchon's 'spectral poetics;' both in its sensitivity to the ghostly realm as well as in its concept of history as a infinitely complex vectore-space in which the movements of individuals can be defined as 'eigenmovements' within a complex spatio-historical matrix. Extending the conceit even further, when Prof. Vanderjuice mentions, on the same page, "Hilbert's recent work on Eigenheit theory" (324), Pynchon ties the mathematical theory of eigenvalues to the political realm *via* Stirner's anarchist philosophy, which is referenced during a conversation of the Belgian nihilists Policarpe and Denis. "'Don't mind Denis, he's a Stirnerite,'" Policarpe says, to which Denis replies, "'Anarcho-individualiste, though you are too much of an imbecile to appreciate this distinction'" (528). The tie-in, of course, is that the first part of the second part of Stirner's *The Ego and its Own,* 'Ich,' is called 'die Eigenheit' – while Hilbert's 'recent work on Eigenheit theory' seems itself to be an imaginary reference.[29]

conclusion: complication

What is the value of this exercise? And how is it related to the 2+ cultures debate? First of all, I think that the use of the notion of eigenvalues and its various modifications by such different 'cultures' as those of mathematics (Hilbert), physics (von Helmholz, Schrödinger), literary studies (Dilthey), systems theory (von Foerster, Luhmann), biology (Maturana|Varela) philosophy (Hume, Peirce, James, Deleuze|Guattari), and literature (Pynchon), is in itself significant in that it might herald an overall movement away from *essences* to *habits* and from *logics* to *resonances.*

Beyond that, it also shows how a conceptual notion develops different 'habits' in its movements across different cultures, conceptual ecologies and modes of

[29] The 'Pynchonwiki' notes that Eigenheit is "[a] term used in some of David Hilbert's mathematical and logical systems, it appears to have several disputed meanings, including something like 'peculiarities' or 'unique values or characterizations' (eigenheiten)" and it makes the connection to Stirner: "But Eigenheit also means :'Own-ness' or 'Self-Ownership,' a concept of the German individualist-anarchist Max Stirner (Johann Caspar Schmidt), an issue of real concern to Kit, both in his immediate situation vis a vis Scarsdale Vibe, and perhaps also because of Stirner's radical individualist concept of trade union activity" (http://against-the-day.pynchonwiki.com/wiki/index.php?title=ATD_318-335).

thought. In this context, it might be emblematic of Deleuze's proposition for a 2+ cultures logic, which I want to put up for debate in the rest of my paper. For this logic, as Deleuze notes in *What is Philosophy?*, "the rule is that the interfering discipline must proceed with its own methods" (217).

For Deleuze, "thought in its three great forms – art, science, and philosophy" (*Philosophy* 197) form inherently consistent 'surfaces of thought' that they assemble from a fundamental multiplicity. Although they have different agendas, all three "confront . . . chaos, laying out a plane, throwing a plane over chaos" (197). If artists, scientists and philosophers themselves are treated as eigenorganizations or autopoietic systems, then 'doing' philosophy, art or science means, in the most general sense, to 'perform habitual cognitive routines that organize life.' Although the three forms of thought throw different planes over the chaos, therefore, they are based on the idea that thought emerges from a plane of material multiplicity from which it is 'categorically' separated, but to which, as a radically *embodied* thought, it remains fully immanent. In other words, because all thought is embodied, the 'set' of the processes of life always exceeds the set of 'its' thought: culture is a 'subset' of a nature that Deleuze sees as a productive field that is not *essential* but *eventual* (nature isn't 'natural,' it is machinic). As Deleuze notes in *The Logic of Sense* about the *event*ualization of philosophy: "to reverse Platonism is first and foremost to remove essences and to substitute events [eigenvalues] in their place, as jets of singularities [eigenorganizations]" (Deleuze, *The Logic* 53).

In particular, "from sentences or their equivalent, philosophy extracts *concepts* (which must not be confused with general or abstract ideas), whereas science extracts *prospects* (propositions that must not be confused with judgements), and art extracts *percepts and affects* (which must not be confused with perceptions and feelings)" (*Philosophy* 24). In this differentiation, the history of the migrations of the term eigenvalue can show how it is used *conceptually* (von Foerster, Dilthey), *prospectually* (Hilbert, Schrödinger, Maturana| and Varela) *and perceptually|affectually* (Pynchon). As Deleuze notes,

> philosophical concepts have events for consistency whereas scientific functions have states of affairs or mixtures for reference: through concepts, philosophy continually extracts a consistent event from states of affairs—a smile without the cat, as it were—whereas through functions, science continually actualizes the event in a state of affairs, thing, or body that can be referred to. (126)

Philosophy "brings forth events. Art erects monuments with its sensations. Science constructs states of affairs with its functions" (199).

Philosophy should never be strictly functional, therefore, nor should a work of art be strictly philosophical, although the three fields are permeable. In the relation between philosophy and science, for example, Deleuze and Guattari ask

> How are we to conceive of practical transitions . . . ? But, above all, theoretically, do the heads of opposition rule out any uniformization and even any reduction of concepts to functives, or the other way around? And it no reduction is possible, how can we think a set of positive relations between the two? (133)

Symptomatically, even Deleuze|Guattari's concept of 'concept' resonates strongly with the scientific theories of autopoiesis and emergence. Such resonances, which also define their repeated references to biology, chemistry and the field of non-linear dynamics, invariably imply a 'becoming philosophy' of science, while their numerous references to literature, painting or film invariably imply a general 'becoming philosophy' of art. In a similar vein, in the work of Michel Serres both philosophy and art are 'becoming science.' Despite the fact that each assemblage has its own logic, and despite the general 'heterogenesis' (199) of thought, therefore, a

> rich tissue of correspondence can be established between the planes . . . the network has its culminating points, where sensation itself becomes sensation of concept or function, where the concept becomes concept of function or of sensation, and where the function becomes function of sensation or concept. (199)

Ultimately, although the three forms of thought form different habitual systems, they operate in a resonant field. All three rely on a field of multiplicity from which eigenvalues|eigenorganizations are generated through recursive operations that create harmonic, resonant systems. As von Foerster notes,

> Wollen wir nun das Problem einer Theorie der Erkenntnis lösen, also eine Epistemologie erzeugen, dann muss sie von solcher Art sein, dass sie sich selbst erklärt, oder in Hilberts Sprache, dass sie eine Eigentheorie ist . . . . Erfahrung ist die Ursache | Die Welt ist die Folge | Die Epistemologie ist die Transformationsregel. (*Wissen* 368-369)

As eigentheories, the three forms of thought create not only different habitual systems but also different epistemologies, with varying relations to the multiplicity from which they emerge: a conceptual, a prospective and a perceptual one. "[P]hilosophy wants to save the infinite by giving it consistency . . . . Science, on the other hand, relinquishes the infinite in order to gain reference . . . . Art wants to create the finite that restores the infinite" (*Philosophy* 197). Dilthey is completely correct, therefore, to state that the hard sciences have a different eigenvalue than the humanities: "Es gibt keine Wissenschaft, welche aneinander binden könnte diese beiden Seiten des menschlichen Wissens: Wissen als Erkennen und Wissen als Erleben und Verstehen" (Dilthey *XX,* 327).

Although the three forms of thought are eigentheories with specific eigenvalues and eigenfrequencies, their ecology relies on resonances that lie at their systemic base. The term eigenvalue, which is itself intimately related to the concept of resonance, attunes artificial crickets that chirp, bridges that fall down, writers that express the intensity of their lives and dentists that drill for historical truths. I would like to hope that the story of the migration of the term has shown some of these resonances.

Works Cited

Braun, Michael. *Differential Equations and Their Applications.* New York: Springer, 1983.

Bateson, Gregory. *Mind and Nature: A Necessary Unity.* Cresskill, NJ: Hampton Press, 2002.

---. *A Sacred Unity: Further Steps to an Ecology of Mind.* London: HarperCollins, 1991.

Calvin, William H. *The Cerebral Code: Thinking a Thought in the Mosaics of the Mind.* Cambridge, MA: MIT Press, 1996.

Cramer, Friedrich. *Symphonie des Lebendigen: Versuch einer allgemeinen Resonanztheorie.* Frankfurt a.M.: Insel, 1996.

De Landa, Manuel. *A Thousand Years of Nonlinear History.* New York: Swerve, 1997.

Deleuze, Gilles. *Cinema 2: The Time-Image.* Minneapolis, MN: University of Minnesota Press, 1998.

---. *Difference and Repetition.* London: Athlone Press, 1994.

---. *Empiricism and Subjectivity: An Essay on Hume's Theory of Human Nature.* New York: Columbia University Press, 1991.

---. *The Logic of Sense.* New York: Columbia University Press, 1990.

Deleuze, Gilles and Felix Guattari. *What is Philosophy?* New York: Columbia University Press, 1994.

Deleuze, Gilles and Claire Parnet. *Dialogues.* New York: Columbia University Press, 1987.

Dilthey, Wilhelm. *Selected Works. Volume III: The Formation of the Historical World in the Human Sciences.* Eds. R. A. Makkreel and Frithjof Rodi. Princeton, NJ: Princeton University Press, 2002.

---. "Das Erleben und die Selbstbiographie (1906-1911/1927)." *Die Autobiographie: Zu Form und Geschichte einer literarischen Gattung.* Ed. Günter Niggl. Darmstadt: Wissenschaftliche Buchgesellschaft, 1989. 21-32.

Dupuy, Jean-Pierre. "Philosophy and Cognition: Historical Roots." *Naturalizing Phenomenology: Issues in Contemporary Phenomenology and Cognitive Sci-*

ence. Eds. Jean Petitot, Francisco J. Varela, Bernard Pachoud, and Jean-Michel Roy. Standford, CA: Stanford University Press, 1999. 539-558.

Frege, Gottlob. *Funktion, Begriff, Bedeutung: Fünf logische Studien*. Göttingen: Vandenhoeck & Ruprecht, 1994.

Friederichs, Karl. *Ökologie als Wissenschaft von der Natur oder biologische Raumforschung*. Leipzig: J.A.Barth, 1937.

Groethuysen, Bernhard. "Anmerkungen." *Wilhelm Diltheys Gesammelte Schriften, VII Band: Der Aufbau der Geschichtlichen Welt in den Geisteswissenschaften*. Leipzig: Teubner, 1927.

Hilbert, David. "Grundzüge einer allgemeinen Theorie der linaren Integralrechnungen (Erste Mitteilung)." *Nachrichten von der Gesellschaft der Wissenschaften zu Göttingen, Mathematisch-Physikalische Klasse, 1.-6. Note*, 1904. 49-91.

James, William. "Psychology: The Briefer Course." Vol. 12. *The Works of William James. 17 Vols*. Eds. Frederick H. Burkhardt, Fredson Bowers, and Ignas K. Skrupskelis. Cambridge, MA: Harvard University Press, 1984.

---. *The Works of William James Vol. 3: Essays in Radical Empiricism*. Cambridge: Harvard University Press, 1976.

Kant, Immanuel. *Groundwork of the Metaphysics of Morals*. Cambridge, MA: Cambridge University Press, 1997.

Lampert, Jay. *Deleuze and Guattari's Philosophy of History*. London: Continuum, 2006.

Lessing, Hans-Ulrich and Frithjof Rodi (eds). *Wilhelm Dilthey, Gesammelte Schriften. Band XX: Logik und System der Philosophischen Wissenschaften*. Göttingen: Vandenhoek & Ruprecht, 1990.

Luhmann, Niklas. *Die Kunst der Gesellschaft*. Frankfurt a.M.: Suhrkamp, 1997.

---. "Deconstruction as Second-Order Observing." *New Literary History* 24.4 (1993): 763-782.

---. *Die Wissenschaft der Gesellschaft*. Frankfurt a.M.: Suhrkamp, 1992.

Maturana, Humberto and Francisco Varela. *The Tree of Knowledge: The Biological Roots of Human Understanding*. Boston, MA: Shambhala, 1998.

Massumi, Brian. *Parables for the Virtual: Movement, Affect, Sensation*. Durham, NC: Duke University Press, 2002.

Peirce, Charles Sanders. "Design and Chance" [1884]. *Writings of Charles S. Peirce: A Chronological Edition. Volume 6, 1886-1890*. Ed. Edition Project. Indianapolis, IN: Indiana University Press, 1986. 544-554.

---. "A Survey of Pragmaticism." *Collected Papers of Charles Sanders Peirce, Vol. V: Pragmatism and Pragmaticism*. Eds. Charles Hartshorne and Paul Weiss. Cambridge, MA: Belknap Press, 1965. 317-345.

---. "Book III: Critical Logic." *Collected Papers of Charles Sanders Peirce, Vol. II: Elements of Logic*. Eds. Charles Hartshorne and Paul Weiss. Cambridge, MA: Belknap Press, 1965. 387.

---. “Issues of Pragmaticism.” *Collected Papers of Charles Sanders Peirce, Vol. V: Pragmatism and Pragmaticism.* Eds. Charles Hartshorne and Paul Weiss. Cambridge, MA: Belknap Press, 1965. 293-316.

---. “What Pragmatism Is.” *Collected Papers of Charles Sanders Peirce, Vol. V: Pragmatism and Pragmaticism.* Eds. Charles Hartshorne and Paul Weiss. Cambridge, MA: Belknap Press, 1965. 272-292.

---. “Book III: Unpublished Papers.” *Collected Papers of Charles Sanders Peirce, Vol. V: Pragmatism and Pragmaticism.* Eds. Charles Hartshorne and Paul Weiss. Cambridge, MA: Belknap Press, 1960. 333.

Pynchon, Thomas. *Against the Day.* New York: Penguin, 2006.

---. *The Crying of Lot 49.* Philadelphia: Lippincott, 1966.

---. *V.* Philadelphia: Lippincott, 1961.

Segal Lynn. *The Dream of Reality: Heinz von Foerster's Constructivism.* New York: Norton, 1986.

Serres, Michel. *The Birth of Physics.* Manchester: Clinamen Press, 2000.

Stephenson, Neal. *Cryptonomicon.* London: Arrow, 2000.

Stirner, Max. *The Ego and his Own.* New York: Harper & Row, 1971.

von Foerster, Heinz. *Understanding Understanding: Essays on Cybernetics and Cognition.* New York: Springer, 2003.

---. *Wissen und Gewissen: Versuch einer Brücke.* Frankfurt a.M.: Suhrkamp, 1993.

---. “Deconstruction as Second-Order Observing.” *New Literary History, Volume* 24.4 (1993): 763-782.

von Helmholtz, Hermann. *Die Lehre von den Tonempfindungen als Physiologische Grundlage für die Theorie der Musik.* Braunschweig: Friedrich Vieweg, 1863.

Wolfe, Cary. “Lose the Building: Systems Theory, Architecture, and Diller+Scofidio's Blur.” *Postmodern Culture* 16.3 (May 2006) <http://muse.jhu.edu/journals/postmodern_culture/toc/pmc16.3.html>.

The New Alliance of Neuroscience and the Humanities: Interdisciplinarity in the Making

Suzanne Nalbantian

The stage is set for a timely consideration of interdisciplinarity between the arts and sciences. As an advocate and practitioner of interdisciplinarity, which has shaped my own career in the past decade, I have reason to believe that it is at the heart of intellectual pursuit in our age. I have arrived at this conclusion from an in-depth study of literature and neuroscience that I have conducted in my book, *Memory in Literature: From Rousseau to Neuroscience* and in the continued research methodologies that I have acquired and shared in ongoing dialogues with noted neuroscientists in the aftermath of that 2003 publication. My tenets are therefore garnered from actual hands-on experiences of generative exchange in stretching across the borders that traditionally isolate these disciplines.

We are in an era for the integration of disciplines, in what might be called a new age of Enlightenment which is gradually producing thinkers, beyond scholars or philosophers. We all know that intellectual approaches come in waves – *Zeitgeists* according to A.O. Lovejoy or paradigms according to Thomas Kuhn's view of scientific revolutions. We can look back to precedents for such interdisciplinary activity in the original Enlightenment figures of the eighteenth century with the bold, eclectic aims of the master Diderot and to the syncretic aspirations of modernist groups like Bloomsbury in England or Surrealism in France in the early twentieth century.

What is fascinating in our time just over a century later is that the loudest call for such cross-disciplinary exchange with the humanities is coming from a group of eminent scientists. This is not surprising given the fact that this has been called the century of the brain; I would even suggest that an intellectual revolution is emerging from the particular domain of neuroscience. Francis Crick's 1994 book *The Astonishing Hypothesis: The Scientific Search for the Soul* first redirected the study of consciousness from the realm of philosophy and religion to the scientific domain of neural correlates, while leaving open possibilities for conciliatory exploration. Then came the American genetically-oriented biologist E.O. Wilson in 1998 with the mandate for the unity of all knowledge, what he labeled as "consilience."

This was an outright rejection of the cleavage between science and literature proclaimed by C. P. Snow in his "two-cultures" theory of 1959, which incidentally made the erroneous historical statement that "very little of twentieth-century science has been assimilated into twentieth century art" (17).

But beyond theorizing, a number of neuroscientists today are attempting to explain in very concrete terms topics that were traditionally in the domain of the humanities, even to the point of adapting their own established scientific theories to this larger perspective. The topic now known as NCC or the neural correlates of consciousness has become a chief goal of leading researchers. Spearheading this approach has been Pasteur Institute's celebrated neurobiologist Jean-Pierre Changeux. In the late 1990s he proceeded to put his own concept of neuronal man to the test through a heated dialogue about the scientific understanding of subjective experience with the phenomenological philosopher Paul Ricoeur in their published dialogue—a characteristically French *débat*—*Ce qui nous fait penser* or *What Makes Us Think.* Since then, Changeux, like other neuroscientists has expanded his domain to "higher level brain processing," moving from the neuronal to the mental. Jointly with Stanislas Dehaene Changeux has proclaimed that "understanding consciousness has become the ultimate intellectual challenge of the new millennium" (1145). Another leading neuroscientist, Joseph LeDoux has moved beyond his initial study of the emotional brain to what he has called the "synaptic self" in his 2003 book of that title. LeDoux offers a completely new grounding for humanists to consider in their understanding of identity and selfhood even though his own scientific research is largely animal-based. In contending that the self and personality are derived from an arrangement of synaptic connections, LeDoux states that his theory "is an attempt to portray the way the psychological, social, moral, aesthetic, or spiritual self is realized" (3). In an even larger perspective joining scientific knowledge with intellectual history, the neurosychologist Stephen Pinker in his popular book *The Blank Slate* has reconsidered the very topic of human nature in terms of natural selection and the genome. In the more focused area of taxonomies of human memory, the neurocognitive psychologist Endel Tulving, has recently revised his pioneering 1972 psychological definition of episodic or autobiographical memory by calling attention to the neurocognitive framework of what he has called "autonoetic awareness," the sense of subjective time and consciousness of self. In a similar vein, the neurologist Antonio Damasio, famous for his earlier book *Descartes' Error* has recently defined the concept of *mind time* as "how we experience the passage of time and how we organize chronology" (68). This kind of evolving orientation that is marking the works of certain well-known scientists shows a concerted effort on their part to avoid reductionism in their theories as they reach out to the issues of the humanists. But humanists have not sufficiently responded to such calls for interdisciplinary dialogue.

Among scientists, meanwhile, the most established interdisciplinary field is the rapidly emerging one of "neuroaesthetics," which received its formal definition in 2002 and has its headquarters in institutes at University College London and at the University of California, Berkeley in the United States. It is construed as the study of the neurobiological bases of aesthetic experience with goals of acquiring a scientific understanding of the contemplation and creation of a work of art. It has also gained ground in the work of certain cognitive psychologists such as Mark Turner and Merlin Donald who have been trying to understand the integrative experience of artistic activity. As leaders in the field of neuroaesthetics, the neurologist Semi Zeki and Jean-Pierre Changeux have studied the neurological underpinnings of the work of art, with Zeki focusing on visual mechanisms. The far-reaching implications of this kind of study are suggested by Zeki's pronouncement that "No theory of aesthetics is likely to be complete, let alone profound, unless it is based on an understanding of the workings of the brain" (103) and by Changeux's affirmation that "the evolution of art . . . incorporates, assimilates and often is inspired by the *données* of science" (*Raison et Plaisir* 143).[1]

It is a fact that a whole variety of "neuro" fields are emerging internationally in relation to all major disciplines from economics to music. Fernando Vidal, a history of science researcher at the Max Planck Institute in Berlin, has pointed out that the numerous "neuro" domains "have tended to promote interdisciplinarity and dialogue between the disciplines" (38).[2] In the overarching field of neuroethics, the University of Pennsylvania psychologist Martha J. Farah has mapped out the variety of ethical issues that advances in neuroscience have raised. Linking brain science to economics, the Princeton University psychologist Jonathan Cohen has been analyzing the impact of emotions upon economic decision-making, what can also be substantiated in part by neuroimaging studies of the prefrontal cortex and its executive control. In yet another domain, the study of the neurobiological foundations of music has been led by the Canadian neurologist group of Isabelle Peretz and Robert Zatorre which is considering music as a window onto complex brain functions, for the understanding of "the interactions between neocortically mediated cognitive processes and subcortically mediated affective responses" (10). At the same time, more traditional subjectve notions like that of *qualia*, signifying the subjective experiences involved in perception, like the redness of a red object, have preoccupied the neural Darwinist Gerald Edelman along with Christof Koch and the late Francis Crick who called it "the most difficult aspect of conscious-

1 Translation from the French Changeux quotation "l'évolution de l'art . . . incorpore, assimile, et souvent s'inspire, des données de la science."

2 Translation from the French Vidal quotation "[I]ls tendent à rechercher l'interdisciplinarité et le dialogue entre les disciplines" (38).

ness"(1133). Modeling subjective experience has become a challenge of its own in all these venues.

Underlying such inquires is of course the entire mind/body issue and its implicit framework of dualism which marks the history of Western philosophy from Plato to Descartes, Kant and Bergson. It is becoming increasingly apparent that this fundamental philosophical discussion is moving from the premises of philosophy departments to those of neuroscience. Brain/Mind inquiries are becoming most prominent among neuroscientists, as they tackle interwoven subjects of cognition and emotion. The once forbidden word "mind" has become rampant as neuroscientists expand and revise their theories to probe higher order brain functions. On this scientific side, this perennial issue first showed up in the consideration of the emotional brain, a subject treated by Antonio Damasio, Edmund Rolls and also Joseph LeDoux. Emotion, traditionally associated with literature, the arts or psychology, has become a special subject for scientists studying its neural and chemical bases. It was Antonio Damasio who in 1994 led the attack on Descartes notion of the disembodied mind in focusing on the somatic consciousness of the body and the neurobiology of rationality. His landmark case study was that of Phineas Gage, the nineteenth-century American construction worker who suffered left frontal lobe damage in an accident that impaired decision-making specifically because emotion and feelings were components that were also compromised. Damasio was thereby attacking the traditional brain/mind dichotomy that also separated reason from emotion. Subsequently Edmund Rolls and Joseph LeDoux initiated the study of emotions as brain functions and beyond. Rolls studied the memory of emotion and its connection to the hippocampus; LeDoux emotional memory, and the model emotion of fear with its ties with the amygdala. Imagine the results if such scientists tapped into the entire reservoir of romantic and modern literature, which portrays a whole range of emotions in the lived experience of humans!

To be sure, such diverse studies offer serious challenges to a whole heritage of literary criticism, as well as to the field of aesthetics, philosophy, psychoanalysis, and finally literary theory itself. If science is seeking enhancement by recourse to the humanities, it also follows that humanities can discover foundations in common with science. Mapping the pathways for twenty-first century literary criticism *The New York Times Magazine* in 2002 stated that "science now holds the glamour spot in academia" and that "Derrida is out." With this attitude, science could become a basis for literary theory. But care must be taken if this to become more than a passing fad. For example, the pitfalls of Structuralism, with the introduction of the social sciences into the study of literature in the 1960s, must be avoided. What began as an interdisciplinary venture in many cases yielded cryptic textual study that ironically isolated the literary texts in closed systems and hermetic formalism.

In our time, under the very noticable impact of the neurosciences, an intellectual triage can begin to take place for students and professors alike in efforts to shed impressionistic and ideological criticism. Most notably, the older psychoanalytical approach to the study of literature has in fact waned in recent years, as the discipline of psychology has become much more of a hard science. That change in orientation has been revealed in the ongoing public debate in the U.S., instigated in the 1990s by the American literary critic Frederick Crews about the validity of the recovered memory movement and the questionable Freudian assumption upon which it is based. So much so that in March 2006 a widely publicized solicitation in newspapers and on the web went out from the Biological Psychiatry Laboratory at McLean Hospital and the Department of Psychiatry at Harvard Medical School for literary documentation as a testing device in the current controversy over the validity of the theory of dissociative amnesia or repressed memory (see Carey and Pope). Despite the fact that no evidence has surfaced to date[3] to deny the hypothesis that repressed memory is a culture-bound syndrome, this heated interchange demonstrates that science is reaching out actively for information from the human experiential reservoir of the humanities to supplement the inconclusive empirical data of the scientific world. The benefits of this public controversy are that it is provoking cross-disciplinary discussion while also demonstrating the need of further scientific evidence for a theory and an approach which once governed an entire sector of literary criticism, as well as psychoanalytic practice.

Moreover, the longtime, still practiced sociological criticism that sees literature as a mimetic reflection of reality, grounded in Eric Auerbach and carried on by a line of contextual critics, would be called "naïve realism" by some scientists. This type of criticism instantly becomes problematic in light of the intricacies of cognitive perception theory moving from the individual brain to the collective consciousness of societies and cultures. Questions about the veridicality of sociocultural memory have been raised, for example, by the American historian Michael Kammen who has stated that "collective memory in the United States has been subject to distortion and alteration" (329). But this issue can be best analyzed in the "micro" neuroscientific framework of *confabulation*—or memory distortion and falsification—which has overtaken a branch of scientific psychology under the leadership of Daniel Schacter, Elizabeth Loftus, Morris Moscovitch and Tim Shallice. In his book *The Seven Sins of Memory* Schacter accounts for the distortion of personal memory by seven kinds of interference including misattribution, suggestibility and bias. The full range of Schacter's work on this topic is a provocative frame of reference that can be added to sharpen literary reception theory as well as historical and cultural criticism. Such a scientific psychologist would be as valuable as any major literary critic precisely because he provides a new gauge for evaluation and analysis.

[3] As of August 23, 2007.

My own focused research on the subject of memory in literature has reconsidered such seminal writers as Proust, Joyce, Woolf, and Faulkner in an entirely new neuroscientific perspective. My treatment of these and other major authors has been informed by the explosion of information about the brain in the 1990s, when the field of neuroscience offered different approaches to understanding memory processes, devised taxonomies for the classification of memory, and provided brain mapping with fMRIs for localising various sites in the brain where some activities take place. I have found that neuroscientific hypotheses regarding the emotional brain, sensory-trigger mechanism, memory traces, the thorny issue of confabulation, the faddish one of reactivated memory and that intricate subject of implicit or unconscious memory can be further explored and understood through the literary expression of modernist memory. In considering such approaches, we, in literary analysis, can also get closer to understanding the creative process which results fundamentally from brain functions that can be partially identified.

Against the fertile background of modern neuroscience, modern autobiographical literature can be used as a virtual laboratory for the study of the workings of the mind in its encoding, storage and retrieval of episodic memory. Using data gleaned from selected literary works to establish correlations with neuroscientific theories and models, I have categorized my findings which corroborate with established taxonomies in neuroscience. Whereas most scientific case studies are based on the impaired or defective individual, my literary case studies are drawn from the healthy, human brain. I have demonstrated that major autobiographical writers of the twentieth century—from Marcel Proust to Octavio Paz—are extraordinary subjects for the study of memory because of their heightened sensitivity, acute perception and astute ability to communicate memory experiences. As outliers, their experiences amplify and heighten those of the normal human brain and therefore can shed light on standard memory functioning. Since most of these works are metacognitive and process-oriented, they actually enact in prescient manner distinct memory processes that are being studied and classified by scientists today. These processes include the different functions of brain structures and the kinds of interconnections from molecular to system function that come into play. Vivid illustrations emerge from these authors.

One of the most sophisticated writers in modern literature can shed light on the most localized and primitive memory processing. Proust demonstrates for us the supreme case of encoded sensory memory mediated in the medial temporal lobe, in what is considered phylogenetically to be the oldest part of the brain. In the case of the famous madeleine episode in the first volume of *A la Recherche du temps perdu* Proust made it clear that it was the olfactory and gustatory senses, taste and smell alone, and not sight, that were the gateways to the vast structure of recollection. For Proust, the limbic material is at the service of his conscious art-making which is triggered by the numerous sensory "reminiscences" in his multivolume opus. In contrast, Virginia Woolf offers the most comprehensive view of the memory process in the traditional encoding, storage, and retrieval phases. Her novel, *To the Lighthouse* with its emphasis on memory in its visual and topographical modality implicates a large range of higher-level cognitive brain activity which coincides with multiple theories about the hippocampal integrative process and its coordination with the neocortex.[4] James Joyce and William Faulkner can be brought into the same context of associative processing which relies on cued retrieval as set forth by Endel Tulving's longstanding descriptions of episodic memory. If Joyce's memory is persistently fashioned throughout his literary opus by auditory and environmental cues of place and locale of his native Ireland, Faulkner's memory derives from different temporal factors which link the personal to the collective past of his native American South.

In contrast to these writers who fall into the category of conscious, declarative memory, other writers and artists can contribute to the understanding of the unconscious and the category of implicit, nondeclarative memory. For example, a modern autobiographical writer like Anaïs Nin sheds light on deeply embedded traumatic memory, which has been presumed by scientists to be modulated by the brain structure of the amygdala. With her striking novelette *Seduction of the Minotaur*, in which the frightening image of a minotaur represents her ambivalent memory of her father, she falls in line with the view of James McGaugh, for example, that "stress hormones released by emotional experiences influence memory consolidation and that the influence is mediated by activation of the amygdala" (17). A surrealist painter like Salvador Dali, on the other hand, who was also the author of an autobiography *The Secret Life* provides in his famous painting *The Persistence of Memory* a visual presentation of what is being called reactivated memory. His 1931 painting demonstrates in a self-reflexive manner a kind of reconsolidation of memory that is conceived to be labile and ever changeable. Moreover, phenomena of "altered consciousness" (1378) that scientists like J. Allan Hobson and Robert Stickgold have tracked as chaotic imagery and bizarre discontinuity in their studies of dreaming correspond directly with those analyzed by literary critics in surrealist poetry. The unanticipated coupling of retrieved images and objects in a poem like Robert Desnos' "Erotisme de la mémoire" illustrates oneiric memory. Finally, poetic language itself in the works of the Latin-American poet Octavio Paz, for example, can be shown to be a conduit for the memory process in a connectionist framework that points to language functions of the brain. The literary manifestations of such language phenomena in an autobiographical poetic work like *Pasado*

[4] For example,Woolf's novel corroborates Sidney Wiener's, Alain Berthoz's, and Michael Zugaro's view that "the hippocampal system receives inputs from multisensory and high order sensory and motor cortical areas and integrates them with inputs concerning emotion and motivation" (75).

en claro corroborates with a prominent neuroscientific view of reconstructive memory derived from the neural network modeling of the neuropsychologist, James McClelland. McClelland's memory trace synthesis model construes memory not simply as recall but as reconstruction often dependent on the strength of connections between units known as connection weights. In light of such literary data, of which this is only a small sampling, I have arrived at a hypothesis that the artistic modification involved in the literary rendering of memory demonstrates the reconstructionist component inherent in the biological workings of memory itself.

Although the particular discipline of comparative literature seems a natural forum for studies of the intriguing connections between literature and neuroscience, historically speaking some comparatists such as Peter Brooks and Francesco Loriggio pointed to the difficulty of such pursuits and the stigma of amateurism that has weakened them. Back in 1989 the outspoken literary critic Stanley Fish came out saying "being interdisciplinary is so very hard to do" (15) because of the power plays involved in annexations of other disciplines. During this period, isolated scholars including Betty Jean Craig and Linda Henderson were among the few who embraced physics in their consideration of art and literature. Craig tackled relativism in literature; Henderson fourth-dimension physics in modern art. But there were few followers, because physics did not have widespread influence among critics. Instead, critics chose other contexts, including cultural studies, gender, race, postcolonialism and transnationalism to globalize literature studies and to question former notions of universality. A new challenge for the future is to link comparative literary study, the most holistic of disciplines, with the hard sciences, which can provide a factual dimension in the understanding of humanity and the natural world. In today's society, a role of the intellectual may very well be to mediate between sophisticated literature and the reading public which seeks scientific facts and statistics that are increasingly used to illuminate the human condition.

While in the past, the stigma of amateurism has been a deterrant to solid interdisciplinary studies, that problem can be avoided in our time in light of the dissemination of scientific theory through serious popularization of its contents. That is to say, through the common network of the internet and with accessible books for the general reading public, hard-core scientists are generously spreading their information to the wider audiences, which is helping to promote the new literacy of science in our era. Literary scholars can be inspired from such available material to explore scientific disciplines which would lend authenticity to their interdisciplinary studies. To do so, critics would have to become earnest autodidacts, propelled by intellectual curiosity, mastering other disciplines on their own or actually pursuing advanced degrees in those other fields.

In order to foster this interdisciplinary exchange, it is important to consider issues of methodology required for serious and fruitful dialogue. Guidelines must be established to encourage such research. Caution must be used in both pedagogical and research contexts, so that appropriate rigor is imposed on this kind of activity. Basic elements of the scientific method can be appropriated. At the same time, if model building is emulated, its provisional basis must be acknowledged. Hypotheses can be rejected through empirical testing but never fully proved. The use of case studies can be valid but must be handled with discretion. Admittedly, generalization from single, individual events—no matter how compelling—is always risky and prone to be misleading. On the other hand, in contrast to statistically derived findings from large samples of observations, case study, as a qualitative and selective form of examination, is more appropriate to explore and render how a particular process or phenomenon materializes rather than to explain why. Moving beyond case studies to broad empirical testing is a further stage that is necessary for more systematic and objective assessments of hypotheses.

For humanists and scientists alike, successful communication necessitates a common vocabulary and understanding of concepts in the history of science and approaches to the critical interpretation of the arts. The stimulating metaphor, used by the French philosopher of science Michel Serres, of Hermes, the messenger between different cultures, highlights the kind of *rapprochement* we should be seeking. A good first step is the creation of a "third discourse" with as many value-free terms as possible. The specific language of each discipline can be translated into a common idiom. At the same time, scientists must be made conversant with the interpretative approach to literary and artistic works and its sometime contextual orientation. All too often, scientists are afraid of the written word and unfamiliar with literary traditions and styles which help classify literature and the arts.

Despite the difficulty of the task, certain paths toward the goal of *real* interdisciplinarity can be paved if the terms of discussion are clarified. Currently, statements from the National Science Foundation and National Institutes for Health in the United States suggest that interdisciplinarity is being understood as a pathway to the future. In particular, Brain/Mind/Behavior centers at Harvard, Princeton, and Stanford have taken the initiative in bringing together research in neuroscience and psychology. But for these universities and other organizations in the United States and abroad, the use of the term "interdisciplinary" seems to denote vast interchange between disparate *scientific* disciplines and may be therefore be better categorized as *intra*disciplinary.[5] These and other funding organizations must

5 For example, the NSF's Office of Multidisciplinary Activities currently funds two projects that are actually intradisciplinary: "Computational Science Training for Undergraduates in the Mathematical Sciences" and "Interdisciplinary Training for Undergraduates in Biological and Mathematical Sciences."(See "Awards," National Science Foundation, at http://www.nsf.gov/funding/funding_results.jsp). Likewise, the rubric of the NIH's interdisciplinary initiative incorporates many branches of neuroscientific investigation, as indicated by its "Neuroscience Blueprint Interdisciplinary Center Core Grants," which support the collaborative research of fifteen NIH scientific institutes and centers (See "Overview Information," Neuroscience Inter-

broaden their inclusion of the humanities in such areas. In the meantime, publishers must be encouraged to accept for publication books which reach disparate reading groups. Specialized journals should take the so-called "risk" of publishing occasional interdisciplinary articles. Moreover, in the curricula of major labs, space should be made for the inclusion of a humanistic component that would also enhance discussions of the perennial ethical issues involved with the advancement of science.

Such an agenda could prove useful in the advancement of this kind of syncretic and holistic perspective that scattered researchers are advocating against the fragmentation of knowledge which has occurred in the isolation of the disciplines. The most continuous group efforts in this direction of linking science and the humanities in the United States are to be found in isolated humanities centers at some universities. However, the institutional context of interdisciplinary study is spotty at best. It does appear that France revealed the potential for the success of such a venture on a huge scale with its popular "l'Université de tous les Savoirs" initiated in the year 2000 to create an ongoing public engagement between the scientific and intellectual worlds—although unfortunately that promising project has dwindled in scope. As a whole, Western culture has not yet reached the kind of integrated, coordinated search for knowledge that would naturally endow researchers with true interdisciplinary orientations. And the competitive monetary incentives of specialization in all fields are in fact deterrants to the fruition of this goal.

The prospective alliance of the humanities and the neurosciences is not just a passing trend or a pipe-dream. Given the crisis of the humanities in our age, a viable means for their survival could very well be their alliance with the sciences. A natural and fruitful start of this intellectual development would be bridging to neuroscience, which is the particular branch of science that explores the distinctly *human* commonalities such as emotion, reason, memory, consciousness, and so on. If such coordinated effort is broadly attempted, literary critics would have to adopt the characteristic humility of the scientist, whose hypotheses are notably open to continuous testing, revision, and debate. The intellectual habit of individual research must yield in part to the collaborative approach of the sciences which gives opportunity for the true meeting of minds and acknowledges the historical evolution of ideas and discoveries. Most responsible scientists admit that if Einstein or Watson and Crick hadn't come up with their discoveries, someone else would have done so at that time. Teamwork must be encouraged for humanists who are ironically often isolated in competitive scheming for theory making and star tracking instead of constructive pathbreaking of knowledge. On their side, scientists must become more verbal and receptive to the imaginative and creative domains of the

disciplinary Center, National Institutes of Health, at http://grants.nih.gov/grants/guide/rfa-files/RFA-NS-06-003.html#PartI).

humanist, outside of the laboratories and on the testing ground of the human, lived experience. Moreover, as science is identifying human universals, it also follows that a global approach must be taken by scientists and humanists to search together for an ethical substratum of human values and behavior.

It is time to seize the inspirational momentum of interdisciplinarity in the making. Such constructive cooperation might well lead to a "one culture" orientation which universities and research institutions world-wide can embrace for the *Wissenschaft* of the future.

Works Cited

Carey, Benedict. "A Study of Memory Looks at Fact and Fiction." *New York Times* 3 February 2007.

Changeux, Jean-Pierre. *Raison et plaisir*. Paris: Editions O. Jacob, 1994

Changeux, Jean-Pierre and Paul Ricoeur. *Ce qui nous fait penser: la nature et la règle*. Paris: Odile Jacob, 1998. [*What Makes Us Think*. Trans. M. B. DeBevoise. Princeton, NJ: Princeton University Press, 2000].

Cohen, Jonathan D. "The Vulcanization of the Human Brain: A Neural Perspective on Interactions between Cognition and Emotion." *The Journal of Economic Perspectives*. 19 (Fall 2005): 3-24.

Craig, Betty Jean (ed.). *Relativism in the Arts*. Athens, GA: The University of Georgia Press, 1983.

Crews, Frederick. *The Memory Wars: Freud's Legacy in Dispute*. New York: New York Review of Books, 1995.

Crick, Francis. *The Astonishing Hypothesis: The Scientific Search for the Soul.* New York: Scribner, 1994.

Crick, Francis and Christof Koch. "A Framework for Consciousness." *The Cognitive Neurosciences*. Ed. Michael S. Gazzaniga. Cambridge, MA: MIT Press, 1995. 1133-1143.

Dalí, Salvador. *The Secret Life of Salvador Dali*. (1942). New York: Dover Publications, 1993.

Damasio, Antonio. "Remembering When." *Scientific American* 287 (Sept. 2002): 66-73.

---. *Descartes' Error: Emotion, Reason, and the Human Brain*. New York: G.P. Putnam's Sons, 1994.

Dehaene, Stanislas and Jean-Pierre Changeux. "Neural Mechanisms for Access to Consciousness." *The Cognitive Neurosciences III*. Ed. Michael S. Gazzaniga. Cambridge, MA: MIT Press, 2004. 1145-1157.

Desnos, Robert. "Érotisme de la mémoire" (1942). *Oeuvres*. Paris: Gallimard, 1999. 813.

Farah, Martha J. "Emerging Ethical Issues in Neuroscience." *Nature Neuroscience* 5 (April 2004): 1123-1130.
Fish, Stanley. "Being Interdisciplinary Is So Very Hard to Do." *Profession* 89 (1989): 15-22.
Henderson, Linda Dalrymple. *The Fourth Dimension and Non-Euclideian Geometry in Modern Art.* Princeton, NJ: Princeton University Press, 1983.
Hobson, J. Alan and Robert Stickgold. "The Conscious State Paradigm: A Neurocognitive Approach to Waking, Sleeping and Dreaming." *The Cognitive Neurosciences.* Ed. Michael S. Gazzaniga. Cambridge, MA: MIT Press, 1995. 1373-1389.
Kammen, Michael. "Some Patterns and Meanings of Memory Distortion in American History." *Memory Distortion.* Ed. Daniel Schacter. Cambridge: Harvard University Press, 1995.
LeDoux, Joseph. *Synaptic Self: How our Brains Become Who We Are.* New York: Penguin; 2002, 93.
---. *The Emotional Brain.* New York: Touchstone, 1996.
Max, D.T. "The Year in Ideas: Darwinian Literary Criticism." *The New York Times* 15 (December 2002): 83.
McClelland, James. "Constructive Memory and Memory Distortions: A Parallel-Distributed Processing Approach." *Memory Distortion.* Ed. Daniel L. Schacter. Cambridge, MA: Harvard University Press, 1995. 69-91.
McGaugh, James L. "The Amygdala Modulates the Consolidation of Memories of Emotionally Arousing Experiences." *Annual Review of Neuroscience* 27 (2004):1-28.
Nalbantian, Suzanne. *Memory in Literature: From Rousseau to Neuroscience.* London and New York: Palgrave/Macmillan and St. Martin's Press, 2003.
---. *Aesthetic Autobiography: From Life to Art in Marcel Proust, Virginia Woolf, James Joyce, and Anaïs Nin.* London: Macmillan, 1994.
Nin, Anaïs. *Seduction of the Minotaur.* Chicago: Swallow Press, 1961.
Paz, Octavio. *A Draft of Shadows and Other Poems.* Ed. and trans. Eliot Weinberger, Elizabeth Bishop and Mark Strand. New York: New Directions, 1972.
Peretz, Isabelle and Robert J. Zatorre. "Brain Organization for Music Processing." *Annual Review of Psychology* 56 (February 2005): 89-114.
Pinker, Stephen. *The Blank Slate: The Modern Denial of Human Nature.* New York: Viking Penguin, 2002.
Pope, Harrison, Jr. and James I. Hudson. "The Repression Challenge." The Biological Psychiatry Laboratory at McLean Hospital <http://biopsychlab.com/chall-en.html>.
Proust. Marcel. *A la Recherche du temps perdu. I. Du Côté de chez Swann.* Paris: Gallimard, 1987.

"Qu'est ce que l'Université de tous les savoirs?" *La webtélévision de l'enseignement supérieur et de la recherche, Canal* <http://www.canal-u.fr/canalu/chainev2/utls/page/chaine/>.

Rolls, Edmund. *The Brain and Emotion*. Oxford: Oxford University Press, 1999.

Schacter, Daniel L. *The Seven Sins of Memory*. New York: Houghton Mifflin, 2001.

Serres, Michel and Bruno Latour. *Conversations on Science, Culture, and Time.*1990. Trans. Roxanne Lapidus. Ann Arbor, MI: University of Michigan Press, 1995.

Snow, C. P. *The Two Cultures and the Scientific Revolution*. New York: Cambridge University Press, 1963.

Tulving, Endel. "Episodic Memory: From Mind to Brain." *Annual Review of Psychology* 53 (2002): 1-25.

---. *Elements of Episodic Memory*. New York: Oxford University Press, 1972.

Vidal, Fernando. "Le Sujet cérébrale: une equisse historique et conceptuelle." *PSN* 3 (January-February 2005): 37-48.

Weiner, Sidney I., Alain Berthoz, and Michaël B. Zugaro. "Multisensory Processing in the Elaboration of Place and Head Direction Responses by Limbic System Neurons." *Cognitive Brain Research* 14 (2002): 75-90.

Wilson, Edward O. *Consilience*. New York: Knopf, 1998.

Woolf, Virginia. *To the Lighthouse*. (1927); New York: Harcourt Brace Jovanovich; 1955: 63.

Zeki, Semir. "Art and the Brain." *Daedalus* 127 (Spring 1998): 71-103 <http://www.neuroesthetics org /research/pdf/Daedalus.pdf>.

All Over Writing: The Electronic Book Review (version 4.0)

Joseph Tabbi

Introducing a collection of scholarly essays, *Debating World Literature*, Christopher Prendergast cites an observation by Arjun Appadurai that should give pause to anyone who wants to create a space for literature in new media: "public spheres," Appadurai writes, are "increasingly dominated by electronic media (and thus delinked from the capacity to read and write)" (22). That "thus" can rankle. Obviously Appadurai is not thinking of the Internet, which is still (and likely always to be[1]) overwhelmingly textual, despite an increasing visual and insistently instrumental presence. The assumption that reading and writing are of course "delinked" from all electronic media, shows just how deep the separation of spheres has become for scholars in the field of post-colonial cultural studies. Any notion that electronic literature might in fact *be* an emerging world literature is foreclosed at the start.

It wasn't supposed to be like this. Appadurai's casual dismissal of reading and writing as active elements in "electronic media" should seem strange, if one recalls the idea advanced by cyberculture visionaries for a universally accessible, open-ended archive primarily for *texts.* That was the idea behind Vannevar Bush's *Memex* and Ted Nelson's "hypertext"—not the current expanse of decontextualized "hot links" but rather a way of bringing documents, in part or in their entirety, to a single writing space for further commentary and the development of conceptual connections. Another word Nelson coined for the process was "transclusion"—an inclusion through site transfers that could be full or partial, depending on one's requirements: in every case, the "original" document remains at its home address while being reproduced at the target address (not just referenced or linked). The achievement of this capacity, which can make reading and researching also a kind of worldwide consortium building, brings to the public activities that had been

[1] Handling images is still something of a strong-man act, at least in applications that I use in my own writing life – which is non-extreme but I think not unrepresentative, for literary scholars with some investment in e-lit. For example, I went over a year using less than 1% of the capacity on my email account, but then the account reached 50% capacity after I circulated among a few friends, resized photos from a single vacation, in a single day.

considered, like much of print culture, private and secluded.[2] Realizing such a collaborative network in the field of literary scholarship is behind the current version of *Electronic Book Review*.

In this essay, I discuss how the interface might be *made* to work in the transformation of critical writing. Electronic interfacing, as practiced since the implementation of *ebr* 4.0 (early in the year 2007), has a chance to bring a distinctively literary practice back into the operational field of computing and text processing. Connections that over time have become, in print, conceptual and implicit, become explicit and readable not through technical means alone (e.g., the "hot link"), but by the strategic placement of words, sentences, and other semantic elements in every space afforded by the screen. Even the URL of an *ebr* publication says something, not only about the electronic address of an essay, a narrative, or an essay-narrative, but about its content; literary concepts are "tagged" in each essay, and the tags are developed in awareness of keyword and metatag development at affiliated sites throughout the Web.[3] Though possible and, in our view, desirable, transformations in the practice of critical writing are by no means inevitable and they will depend, not any one site, but on the development of a consortium of sites and a consensus about "best practices" that answer to, and can help direct, practices under development.

The creation of the *ebr* writing space, even as it looks back to Ted Nelson, also looks forward to another, as yet unrealized, conception of knowledge processing on the Internet – namely, the Semantic Web. The Semantic Web is most useful as a metaphor at this point, since its realization depends not on a top-down development but on the independent decisions by many site developers to mark up and tag text according to a common and communicating set of references.[4] My interest in the Semantic Web is its potential, through the mundane task of tagging documents, for developing not only a database but a vocabulary specific to the field of e-lit, using procedures that involve both ordinary readers and editor/curators. At the same time, the necessary awareness of vocabularies under development elsewhere must not influence the autonomous development of a metatag vocabulary for literary purposes. In whatever ways the literary field is transformed by electronic environ-

[2] One instance of "all over" textual distribution is self-exemplifying in the present essay: namely, the sentences leading up to this point in the essay also serve as an introduction to a companion essay, "Electronic Literature as World Literature," under consideration for print publication in a special number of *Poetics Today* on the topic, "Writing Under Constraint." Otherwise, there is no overlap between that essay and this one.

[3] Since this essay was presented in the Summer 2007 Munich conference, the *electronic book review* has been included in a consortium of sites gathered by the Electronic Literature Organization, committed to developing standards for peer-to-peer, communitarian reviewing in networked environments. See, www.eliterature.org.

[4] I have described the "Semantic Web Applicability" to literature in an essay at the Electronic Literature Organization website.< http://eliterature.org/publications/>.

ments, its transformations should be readable in terms created by, and for, literary authors.

Certainly, its creators want to make the conceptual writing in *ebr* consistent with the predominant flow of information among sites whose developers recognize the need for pooling content. Nonetheless, editors and authors cannot assume that our attention to "semantic" content will be enough to sustain a literary presence on the Internet. Where tagging and linking depend on direct, imposed connectivity at the level of the signifier, the creation of literary value depends on suggestiveness, associative thought, ambiguity in expression and intent, fuzzy logic, and verbal resonance (where slight differences, not identifications among fixities, are the origin of meaning – "the difference that makes a difference," in Gregory Bateson's phrase; or, in Emily Dickinson's expression of "internal difference / where the meanings are," the "topologies" that, according to Michel Serres in his book, *Atlas*, "haunt" the geometries where most people live). Tags are important; naming is one of the literary arts. But the names need to change, new names need continually to be created so that the tags read by machines do not appear (to living readers) as word soup. The need to combine this literary development with the machine-readable content that would characterize an operative Semantic Web, is a challenge not only for *ebr* but for any site interested in knowledge creation that depends on, but is never identical to, information storage and retrieval.

All Over Writing

At a time when powerful and enforced combinations of image and text threaten to obscure the differential and processual ground of meaning, editors will want to recognize and encourage the potential for bringing together, rather than separating, rhetorical modes in the production of nuanced, textured languages within electronic environments. Much of what the *ebr* editors present, online, is recognizable from the tradition of print: the self-standing essay, the book review, editorials, descriptive blurbs, and so forth. What distinguishes our presentation from print, however, is a way of linking content together through conceptual writing, so that relations that tend to be implicit in a print archive are made explicit and present in one place. Following a reference or an allusion or even a hint, readers needn't go to a different bookshelf, library, or archive. The term I want to offer, for such a critical enterprise, is drawn from the arts: bearing in mind the "all over painting" in abstract expressionism, I want to propose an "all over writing" that embraces seriality and interconnectivity, rather than being distracted by links. It happens that this term, "all over painting," figured in my first book, on relations of technology and contemporary fiction, which was published at about the time when I conceived *The Electronic Book Review*. For documentary purposes, as well as for purposes of visual illustration, I will launch this discussion with a brief reference to my book – or rather, to the cover:

Postmodern Sublime (Ithaca, NY: Cornell University Press, 1995)

Reproduced here is "Small Higher Valley 1" (1991), the first in a series of paintings by the New York based poet and painter Marjorie Welish. In the book's introduction, I described this painting's use of "a virtual system of geometric sectionings to suggest the networks and grids that underlie rational thought," while at the same time avoiding any single total system that could dominate everything *(Postmodern Sublime 20)*. That description resonated with my topic, the sublime in American fiction as it was finding expression in a group of authors whose work registers the emergence, post-World War II, of technologies of information, communication, and control. Only recently, while unpacking my library in the Summer of 2007, did I happen to notice how similar in some ways Welish's multi-colored grids are to the visual design of the *ebr* weave by Anne Burdick (http://www.electronicbookreview.com/).

Notice for example this page's capacity to include any color against the generous black background, its flexibility and expansiveness made possible, not limited, by material constraints of the line and the screen. The grid gives precise and measurable locations – they are the known habitations for both the viewer of Welish's painting, and the user of interfaces. But, to cite Serres again, one can live in geometry and still be haunted by topology. The grids also serve to stage a sequence of wholly *relational* meanings. Paint is allowed to brush or bleed into the adjoining quadrilateral sector (in the Welish painting); while files, placed under columns and listed chronologically, also radiate outward to other files based not only on informational content, but on conceptual similarities that might be recognized by readers and editors, though not necessarily anticipated by authors. The design is relational and open-ended, and the electronic writing space is extended "all over," so that one site or essay can be included within and transferred to other sites.

Market

Here, I want to discuss some of the ways that our mode of "all-over writing" capitalizes on what the Web allows, enabling a media-specific reading and writing practice. But to understand this specificity and why it requires designers and programmers working with, but necessarily independent of, writers, I need to say a few words about what our journal is *not* trying to do. First of all, we're not competing with print, and we're not trying to reproduce the peer review academic journal or the just-in-time delivery of established review media. Though it sounds odd to say it, even slack – there's no reason that an essay or book review needs to appear close in time to the texts under discussion, except for the commercial (and relatively recent) enforcement of brief shelf lives for books on the one hand, and platform obsolescence for most Internet sites on the other hand. Those limitations are not inherent in books or websites. Obsolescence—a theme in *ebr's* earliest manifesto—is not a technical problem, but a political and economic one (cf. Tabbi, "A Review of Books").

With the rise of neo-liberal economics in the1980s and 1990s came a consolidation of major book publishers, a proliferation of small presses even as local bookstores were in decline. Authors were being transformed into performers and book peddlers, and the desktop itself was being converted into an environment suitable primarily for office work and innovations in marketing. In general, such transformations have been catastrophic for reading and writing, as my opening quotation from *Debating World Literature* suggests. The Web's been around long enough, that one can safely say that older, bounded, forms of the literary are not likely to re-appear in the current, user-friendly environments. There is of course a wealth of experimental, non-narrative fiction and poetry in non-commercial platforms. But if we haven't had major born digital novels or poems by now, probably we never will.

In my own practice of critical writing within English and Arts programs in the United States, these developments would seem to be consistent with the rise of cultural studies and media studies – where graduate students, instead of taking the years and sometimes decades needed for mastering a subject, are encouraged to publish even as they are still taking courses, carrying a teaching load, and often holding a job. There are now, at my State University in Chicago, even academic conferences for *undergraduates*. Literary and Cultural journals on the Internet tend to be short-lived, and platform obsolescence has made it difficult to establish the canon of contemporary texts that is necessary to the sustained critical discussions needed to form a field. The literary work produced in such a climate has been characterized, in an essay presented to the German 'Network' on "American Studies as Media Studies" by the critic John Durham Peters, as a kind of "just-in-time production." With low start-up costs and low barriers to entry in terms of knowledge, and the ability to "supply the increasingly important cultural industries with savvy employees," the new curricula can be recognized as a kind of "academic parallel to new-liberal economic policies." Such curricula, and the mostly student-run Internet journals that support them, often encourage a topical, informatic approach to scholarship that might be summarized in the formula: "find a hot topic, add theory, present paper."

If the Internet were just a way of making that process still more efficient, I would have left the field years ago. In fact, all but a few of my colleagues in Literary Criticism, Theory, Fiction and Poetry Writing, have left, or relocated to departments of New Media, Arts, and Communications. After establishing a career where reading, writing, and traveling are the primary activities, what author would compromise such autonomy for a career of Project Management, Grant writing, long-distance and frequent commuting to corporate conferences, and continual subjection to programs and platforms that routinely confuse commercial interruption and technical instruction? As Linda Brigham writes in her *ebr* review of N Katherine Hayles, there is something abject about our dependence on expensive, controlled goods, and even the celebrated distribution of agency in networks has its limits. "Feedback to network nodes," Brigham writes, "seldom indicates the nature of the network; *that* information yields only to a higher level of surveillance and analysis, while the nature of the network feeds some entity beyond us, we continue to subsist on the empty calories of ideas and concepts." A similar note is sounded by Andrew McMurry, in his introduction to our *Critical Ecologies* thread: the idea that *everything*, even literature, needs to be done using computers, might serve the current technocracy but it has mostly rendered transactions and communications "sclerotic."

Unforgiving as this critique may be—and I agree with it—the work of McMurry, Brigham, Peters, and many others writing for *ebr* also indicates a way forward: these writers, after all, are not *only* offering critiques; and neither are they simply transcribing their critical writing from one medium to another. What they are doing, in most cases, is *reflecting on* the medium and their own relation to networks as they join with (or engage in principled argument against) other writers within a network that is identified in the process of writing. This engagement involves more than an adjustment of attitude or achievement of competence with computers and databases. To subsist on more than concepts, one needs to bring one's own work into contact with other, related work, so as to be recognized by others who have made similar recognitions on the basis of involvement in similar projects, similar discussions. Every technical innovation in *ebr*, fundamentally, is geared toward the realization of this one goal: to bring the electronic network and its nature into critical consciousness, in the place where such criticism is produced. What we are working toward is the possibility not simply of literature's inhabiting networks, but for literature to *become* a network.

Emergence

Once that goal is recognized, it becomes possible to imagine a place for doing the work of literature without expecting progressive advancement, revolution, or the end of books. I refer to "the work of literature," not "works" of literature, for a reason, namely: talking about processes makes more sense in electronic environments than talking about objects, even when the objects are verbally inventive and could only be devised using new media. At *ebr*, in the threads titled *Webarts* and *Image + Narrative*, we give extensive coverage to conceptual and literary arts that explore their newfound media specificity, but we're not a free-standing art project. *ebr* accommodates, but does not encourage, critical hypertexts and other self-contained, custom projects because these tend to proliferate connections internally, encouraging reading in isolation.

Knowledge, in such an all-over writing environment, is produced not directly, but as a meta-phenomenon, traceable to (though not identical with) the tags, keywords, and descriptors that authors use (or that they leave to the programs they're using, in which case authors cede more autonomy to the machine than they might know). Here, the actual knowledge is not produced, not entirely, by the content of an essay but is "put in" by the author or editor, and the uses of such knowledge are not realized until a reader enters the picture, following a gloss, or connecting one tag with another, identical or related tag. In this sense, knowledge production in a networked environment is "virtual" – which is to say, it is given as a potential, in the act of tagging, and realized only when the relations among tags are recognized by a reader or made noticeable by an editor.

In following such connections and enfoldings, the reader does not in any sense *replace* the author as a producer of knowledge; rather, the reader produces a different knowledge, constructed not only from works the reader has read, but from the works's self-descriptions. The knowledge is, from the very start, already relational: and this is what makes it appropriate for a networked environment driven as much by semantic encoding (regarding what works are *about*) as they are by syntactic and structural coding (regarding what the works are made of materially – its letters, sentences, and so forth).

Jerome McGann, in his *ebr* essay on the electronic future of the Humanities, mentions in passing the "severe critique of critique from what D. G. Rosetti called 'an inner standing-point' – that most telling of critical positions." McGann of course creates his own set of critical references, including his near-contemporary, Bruno Latour, as well as past self-critical creators, critics, and philosophers such as Rossetti and Alfred North Whitehead. That is what any scholar must do, in addressing him or herself to peers in a literary essay. But, in addition to the author's self-chosen references, the gloss on Rosetti takes readers to a critique of McGann's own Rossetti Archive, by Katherine Acheson in *ebr*. Still further, but invisibly, the term, from "an inner standing-point" has been tagged with the keyword, "focalization." And so the entire essay is not just linked notionally to essays on or by Rossetti, McGann, and the field McGann consciously enters; McGann's work also has been gathered, through the database, to literary works by (for example, Rob Swigart's short story, "Dispersion") that experiment with focalization as well as several essays that discuss the concept critically. Further still, once the tag is in place, it will be linked automatically to future works on that topic, as they are recognized and tagged by future *ebr* editors.

Truly, if there is such a thing as "all-over writing," it cannot be defined by pointing to specific features or information; the site needs to be worked with, read, so that significances that are relational have time to emerge. (A similar, serial effect is given in all-over painting: the "Small" and the "Higher" in Welish's "Valleys" would be impossible to discern on just one canvas from this series: smaller, higher, than *what*? *Where*? Such questions are meaningful only with regard to relations that are produced as the painting is created, and as the painter's decisions are recognized by viewers over an extended time of viewing.) Emergence does not produce an object; meaning in online writing cannot be traced or reconstructed by monitoring hits or reader trajectories, meaning can only be held in mind, while reading, writing, or gathering essays onsite. I'll give here two examples – both of them conventional enough to look at, but connected in ways that are recognizable in the process of reading the essay. In doing this, I present *ebr's* first "enfolded" site – a project description at the University of Virginia "NINES" website that is coherent with what's been happening at *ebr*. Rather than simply "linking" to this site, we've brought the essays in their entirety from the NINES site into *ebr*. The essay remains on the NINES site, but its description, its metadata pointers, are brought into the *ebr* database. The essay itself is, in a sense, "wrapped," so that (from a reader's point of view) it is as much a part of *ebr* as the essay by Jerome McGann which mentions the project (a "Networked Infrastructure for Nineteenth-century Electronic Scholarship").

The NINES and COLLEX projects, referenced by McGann in *ebr,* can be accessed on the University of Virginia Web Site as well as at *ebr*. The blurb leading to this essay stresses the coherence (and difference) between MgGann's and Nowisky's project and the overall ('all-over") ambition of the *ebr* interface, namely:

> NINES is an initiative at the University of Virginia to "establish a coordinated network of peer-reviewed content and tools." We present the project here because it's consistent with the initiative at ebr to create a peer-to-peer literary network for conceptual writing.

This example is meant to demonstrate the reach of the *ebr* interface, a model for collaborative reading, and a mode of collaboration among sites. The ease of "linking" makes it unlikely that editors will consider, in detail, what arguments, keywords, metatags, and implied audiences essays from the two sites might have in common. That such collaboration demands explicit negotiation between site editors, who are expected to grant permission without seeking payment in return, is a necessary and desirable feature for the construction of a literary network. It carries into the new media one aspect of scholarly interaction that critics and writers cannot readily do without, namely, the gift economy among literary and cultural peers.

Coda: Peer to Peer

No one, I think, will dispute the desirability of developing Semantic Web standards that are suitable for literature. Few ought to object in principle to establishing consortia of mutually recognized sites so that a vocabulary standard can evolve over time and under a range of institutional contexts. For such cooperation to gain traction, however, mechanisms of review need to evolve along with the bibliographic and semantic standards. These review mechanisms, to be more than privileged community gates, also need to be in place universally, throughout the literary profession. In the past, at universities worldwide, the peer review system has developed in response to this need for standards, which is in reality twofold: 1) to keep track of terminology and conceptual trends so as not to turn the ivory tower into a tower of babble; but also, 2) to uphold standards of quality. The communicative function, which is managerial, is not always conducive to the qualitative function, which is a matter of agreement and disagreement among many subjectivities, among authors, readers, and (most important to the field development) readers in the process of becoming authors. Bringing these two functions together, the administrative and the evaluative, is the challenge of academic review.

Beyond even these dual necessities of quality control and career advancement, peer review is also, perhaps primarily, a mechanism of sorting. It's how professionals, would-be colleagues and collaborators, select some materials for attention and reference (in the process necessarily excluding the majority of materials and producers). I say, "necessarily," because only through selectivity can the efforts of professional readers and their students be responsibly marshaled. Reading lives are limited, and this material condition necessitates the development of literary canons and at the same time justifies the employment of accredited professionals to teach canonical works and their differing receptions in different historical periods.

While necessary in principle, peer review can easily be corrupted when faculty themselves no longer have time collectively to read the work that their own profession is producing at record volumes even as tenured lines at universities worldwide have been reduced drastically. As is widely known and trenchantly reported by Marc Bousquet, in the United States today, around 75% of courses are taught by lecturers, graduate employees, and other casual or temporary workers; 25% by tenured or tenurable professors. Forty years ago the proportions were reversed.

Forty years ago, Roland Barthes could plausibly define the literary "canon," quite simply, as those books that are taught in schools. Today, when what is taught in one humanities classroom is rarely mentioned in another, the canon is reduced to works that get anthologized. In the one class that most undergraduate students have in common, Freshman Composition, literature is often excluded altogether, and students are evaluated according to standardized tests rather than by standards developed intrinsically by the lecturer in collaboration with peers and in response to student feedback. In such circumstances, the transfer of knowledge needed to pass the tests can be readily outsourced to non-tenurable faculty.

The news about job casualization would not be so bad, if Bousquet were simply arguing that the academic system is currently dysfunctional and in need of repair. But Bousquet's more provocative argument is that university administrations, with the implicit support of faculty, have shaped the system to do exactly what it is meant to do – namely, to restrict the supply of peer reviewed researchers and employ an expanding force of low-wage workers whose development is subject to sub-professional performance standards. Consistent with critiques of new liberal economics generally and the critiques of "just-in-time" scholarship by McMurry, Brigham, and Peters in *ebr* (cited above), Bousquet connects the "informatics" of education with the "informality" of work conditions for the majority of graduate and non-tenurable teachers. From this perspective, fears of the traditional university's being displaced by electronic regimes of "distance learning" are misplaced. The packaging of education as information has already "distanced" the majority of literary professionals from the day to day activity of their own students and colleagues even within their own departments. The restrictions on what we actually get to "review" are such that our collective work, as writers and scholars, is unknown even to ourselves.[5]

The consequences of this material transformation are felt in the tenured ranks themselves in many ways. Restrictions on what can be reviewed, and on the number of professors enlisted to do the reviewing, can in effect disqualify professors from evaluating peers responsibly and selecting the texts that will be common to our disciplines. This is not an attack on "academic freedom" as such; the disqualification of the professoriate has more to do with limitations on time and material circumstances (the control of travel and access to grants and other means of connecting with peers). Speaking from my own experience, over the past several years at advisory meetings for the promotion of colleagues, I increasingly have a deep sense that only those assigned to report on a candidate's scholarship have read the work with any care. The majority have time only to read reports sent in by "outside readers"- who in reality are often readers suggested by the candidates under review. The premise, that there is an audience of specialists "out there," better able to analyze a work of literary writing than the colleagues in one's own department, has done more to fragment the profession than any purported tendency toward jargon or politicized language in academic writing. The conceit that there is a set of standards apart from those developed internally among a cohort of professionals, only reinforces the widespread acquiescence to the imposition of standardized testing at all levels of education.

So as to avoid this outsourcing of services that need to be performed by all, not a select few, within the literary profession, and to advance efforts at reforming the academic review process, as of August 2008 *ebr* has placed on its site a formal statement of our longstanding, hitherto informal, practice:

[5] See the "Techno-Capitalism" thread in *ebr*, co-edited by Bousquet and Katherine Wills. The ebr peer to peer model was introduced in 2005, for the launch of the Alt-X/ebr critical e-book series. See Joseph Tabbi, "The Politics of Information: A Critical E-Book Under Way."

> *ebr* is a journal of critical writing produced and published by writers for writers: a peer to peer modification of academic review. Each essay is reviewed by a thread editor (a tenured professor) and at least one other *ebr* editor. On acceptance, the essay is posted to our staging site, where it is made available for comment by our 500-plus past contributors, all of whom are published authors in print and online. Unlike academic peer reports, which are generally seen only by committees, *ebr* reviewer comments can be read in the margins of the essays, as "glosses." More substantial response is given in commissioned Ripostes.

This policy is in solidarity with initiatives and institutional experiments under way as of this writing, notably at the Institute for the Future of the Book (cf. Young).

More generally, the development of a web-based reading culture promises to bring to academia and its publishing institutions something that has been languishing in print culture for a long time, namely: a practice where works are not only read but our various readings are recorded, and that record is itself made public. Like the standardized tests that can account only for what can be tested, most accounts of "the decline of reading" can account only for elements that can be measured: in surveys giving the number of books in circulation, the time that students or teachers claim to spend reading, and so forth. If instead of measuring what is measurable, we make visible the active and participatory reading that is actually going on in our profession, we improve our chances of explaining what we do to those outside the fields of scholarship and creative writing. What we bring to our respective desktops, and what we do with the materials that arrive there, is the essence of literary work. The activitation of this process, and the case by case transclusion of work by our self-selected colleagues, is not just a realization of the technical promise of literary hypertext. The idea is not just to establish digital writing practices as one further literary specialization among all the others. The goal of an all-over writing project has not changed since the work and vision of Ted Nelson forty years ago: to renew literary scholarship as such.

Works Cited

Acheson, Katherine. "Multimedia Textuality; or, an Oxymoron for the Present." *ebr – electronic book review* 11-11-2006
<http://www.electronicbookreview.com/thread/criticalecologies/illuminated>.

Bateson, Gregory. *Steps to an Ecology of Mind*. Boulder, CO: Paladin Books, 1973.

Bousquet, Marc. *How the University Works: Higher Education and the Low-Wage Nation*. New York: New York University Press, 2008.

Brigham, Linda. “Do Androids Dream of Electronic Mothers.” *ebr* 11-09-2006 <http://www.electronicbookreview.com/thread/electropoetics/liminal>.

McGann, Jerome. “The Way We Live Now, What is to be Done.” *ebr* 01-03-2007 <http://www.electronicbookreview.com/thread/electropoetics/rethinking>.

McGann, Jerome and Bethany Nowisky. “NINES: A Federated Model for Integrating Digital Scholarship.” *ebr* 04-09-2007 <http://www.electronicbookreview.com/thread/enfolded/collaborative>.

McMurry, Andrew. “Critical Ecologies: Ten Years Later.” *ebr* 12-01-2006 <http://www.electronicbookreview.com/thread/electropoetics/ecocritical>.

Peters, John Durham. “Strange Sympathies: Horizons of Media Theory in America and Germany.” Paper presented at the annual conference of the German Association of American Studies (Deutsche Gesellschaft für Amerikastudien). “American Studies as Media Studies.” Göttingen, 10 June 2006. Also as “Strange Sympathies” in *ebr*: <http://www.electronicbookreview.com/thread/criticalecologies/myopic>.

Prendergast, Christopher. *Debating World Literature*. Durham: Duke University Press, 2004.

Serres, Michel. *Atlas*. Paris: Flammarion, 1997.

Swigart, Rob. “Dispersion.” *ebr* 10-27-2006 <http://www.electronicbookreview.com/thread/fictionspresent/apparent>.

Tabbi, Joseph. *Postmodern Sublime*. Ithaca, NY: Cornell University Press, 1995.

---. “A Review of Books in the Age of their Technological Obsolescence.” *ebr* 12-30 (1995) <http://www.electronicbookreview.com/thread/electropoetics/manifesto>.

---, “The Politics of Information: A Critical E-Book Under Way.” <http://www.electronicbookreview.com/thread/technocapitalism/editorial>.

Young, Jeffrey R. “Blog Comments and Peer Review Go Head to Head to See Which Makes a Book Better.” *The Chronical of Higher Education*, Tuesday, January 22, 2008 <http://chronicle.com/free/2008/01/1322n.htm (for subscribers only)>.

From Ludwig Boltzmann's Formula to Meatball Mulligan's Party; or How to Fictionalize the Entropy Law

Peter Freese

> *She did gather that there were two distinct kinds of this entropy. One having to do with heat-engines, the other to do with communication.*
>
> Thomas Pynchon, *The Crying of Lot 49* 77

The mysterious notion called 'entropy' already had a highly controversial history when in the 1870s Ludwig Boltzmann made his controversial move from macroscopic thermodynamics to statistical mechanics and, with his famous formula $S = k \log W + c$, replaced the measurement of the vaguely defined property called 'heat' by a stochastic computation of molecular movements as dependent upon changing temperatures. Sadi Carnot's initial hunches in his *Réflexions sur la puissance motrice du feu* (1824) about the unavoidable loss incurred during the transformation of thermal into mechanical energy in steam engines caused William Thomson, the later Lord Kelvin, to observe in 1852 that

> within a finite period of time past the earth must have been, and within a finite period of time to come the earth must again be, unfit for the habitation of man as at present constituted, unless operations have been, or are to be performed, which are impossible under the laws to which the known operations going on at present in the material world are subject. (306)

Two years later, Hermann von Helmholtz confirmed such a prediction by stating:

> . . . dass der erste Theil des Kraftvorraths, die unveränderliche Wärme, bei jedem Naturprocesse fortdauernd zunimmt, der zweite, der der mechanischen, elektrischen, chemischen Kräfte, fortdauernd abnimmt; und wenn das Weltall ungestört dem Ablaufe seiner physikalischen Processe überlassen wird, wird endlich aller Kraftvorrath in Wärme übergehen und alle Wärme in das Gleichgewicht der Temperatur kommen. Dann ist jede Möglichkeit erschöpft, dann muss vollständiger Stillstand aller Naturprocesse von jeder nur möglichen Art eintreten. Auch das Leben der Pflanzen, Menschen und Thiere kann natürlich nicht weiter bestehen, wenn die Sonne ihre höhere Temperatur und damit ihr

> Licht verloren hat, wenn sämmtliche Bestandtheile der Erdoberfläche die chemischen Verbindungen geschlossen haben werden, welche ihre Verwandtschaftskräfte fordern. Kurz das Weltall wird von da an zu ewiger Ruhe verurtheilt sein. (116 f.)

Thus, the notion of the world's irreversible drift towards what came to be called its eventual 'heat-death' entered the scientific discourse, and a misleading term was coined which sundry science-fiction writers would creatively misunderstand as a death *by* instead of a death *of* heat. Many scientists remained skeptical with regard to such a universal *memento mori*. Thus, a pious James Prescott Joule doubted that God had really built an irredeemable flaw into his creation when he said that "we might reason, *à priori*, that such absolute destruction of living force cannot possibly take place because it is manifestly absurd to suppose that the powers with which God has endowed matter can be destroyed any more than they can be created by man's agency" (I, 268 f.). And in a letter of 11 December 1867 to Peter Guthrie Tait, James Clerk Maxwell devised his 'intelligent doorkeeper,' which he then incorporated into his *Theory of Heat* (1871) and which in 1879 Thomson would famously dub 'Maxwell's Demon,' and he did so in order to refute what young Rudolf Clausius had expressed in the two laconic sentences "Die Energie der Welt ist constant" and "Die Entropie der Welt strebt einem Maximum zu" (400). In the ruling climate of steady growth and unchecked progress, in which Darwin's theory of evolution promised the very opposite of entropic dissipation and the aspiring European powers were building their colonial empires, the notion of a scientifically substantiated end of the universe, which—in T. S. Eliot's famous phrase—would come "not with a bang but a whimper" (86), seemed quite unacceptable, and many natural scientists could not accept what Max Planck would later call "the hypothesis of elementary disorder" (50).

In the cultural realm, disbelief was even more persistent. Henri Bergson, who characterized the Second Law as "la plus métaphysique des lois de la physique" (701), invented the notion of *élan vital* as an energy that was exempt from the workings of entropy. In *Der Wille zur Macht: Versuch einer Umwertung aller Werte*, Friedrich Nietzsche devised his theory of eternal return in order to refute what he considered the unacceptable threat of a final state (see 167 f.). Friedrich Engels rejected the Second Law because it ran counter to his belief in the eventual triumph of the proletariat (see 278), and later even Pope Pius XII had a Vatican conference declare that the entropy notion did not invalidate the belief in a primal creator (see Kannegiesser 852). But it was not only the frightening finality and forbidding abstractness of 'entropy' that kept it from becoming common knowledge, but, ironically enough, also its very name. Clausius, who felt that scientific concepts should have Latin or Greek names to be internationally understood, explained:

> Sucht man für S einen bezeichnenden Namen, so könnte man, ähnlich wie von der Größe U gesagt ist, sie sey der *Wärme- und Werkinhalt* des Körpers, von der Größe S sagen, sie sey der *Verwandlungsinhalt* des Körpers. Da ich es aber für besser halte, die Namen derartiger für die Wissenschaft wichtiger Größen aus den alten Sprachen zu entnehmen, damit sie unverändert in allen neuen Sprachen angewandt werden können, so schlage ich vor, die Größe S nach dem griechischen Wort η τροπη, die Verwandlung, die *Entropie* des Körpers zu nennen. Das Wort *Entropie* habe ich absichtlich dem Wort *Energie* möglichst ähnlich gebildet, denn die beiden Größen, welche durch diese Worte benannt werden sollen, sind ihren physikalischen Bedeutungen nach einander so nahe verwandt, daß eine gewisse Gleichartigkeit in der Benennung mir zweckmäßig zu seyn scheint. (390)

Clausius' mistaken notion that *energy* was a compound of *en+ergon* = 'work content' and his coinage of the parallel compound of *en+trope* = 'transformation content' introduced yet another controversial term. Soon Felix Auerbach would suggest that *ectropy* would be a more appropriate term, and later Heinz von Foerster would complain that "we are stuck with the wrong terminology. And what is worse, nobody checked it! An incredible state of affairs!" (181) and, in turn misreading Clausius' etymology, would suggest that the correct term should be *utropy*.

As if this was not complicated enough, in 1948 Claude Shannon, an engineer from the Bell Telephone Laboratories, computed the maximum of messages to be sent through a cable and to his surprise came up with an equation that was identical with that of entropy in statistical mechanics. Before he published what has become the accepted function, he asked John von Neumann for a suitable name for the new property and the latter gave him an advice that is a welcome consolation for puzzled lay persons: "You should call it 'entropy' and for two reasons: first, the function is already in use in thermodynamics under that name; second, and more importantly, most people don't know what entropy really is, and if you use the word 'entropy' in an argument you will win every time" (Tribus 2 f.). With Shannon's statement that "quantities of the form $H = - \sum p_1 \log p_1$ play a central role in information theory as measures of information, choice and uncertainty" and his comment that "the form of H will be recognized as that of entropy as defined in certain formulations of statistical mechanics" (50), the entropy concept entered the newly developing field of information theory.

When Norbert Wiener explained the social consequences of this new discipline to a lay audience in his popular book *The Human Use of Human Beings: Cybernetics and Society* (1950), the notion of informational entropy entered public discourse and made the relation between thermodynamic and informational entropy a hotly contested issue mainly with regard to two aspects. The first derives from Boltzmann's macrostate-microstate distinction, which turned the macroscopic entropy-notion of earlier thermodynamics into a stochastic concept and, by relating entropy to uncertainty, raised the vexing question as to whether this uncertainty refers to the state of the system observed or only to the observer's lack of knowledge about it. Today, there are the two camps of the materialists and the mentalists who conceive of 'entropy' as either the immanent property of a given system or the measurement of an observer's lack of information about this system. The second issue arises from the unanswered question as to whether 'entropy' in information theory and 'entropy' in statistical mechanics are identical or only analogous. Both issues are closely related, since someone who conceives of a gain in entropy as a loss in information and vice versa will more easily concede the identity of the two entropy concepts than someone who insists on the 'objective' quality of thermodynamic entropy and thus sees informational entropy as nothing but a heuristic analogy. Today, there are many scientists who claim that the Boltzmann and the Shannon equations are only symbolically isomorphic but otherwise have little in common.

This was already sufficiently difficult, but with regard to 'entropy' as a concept of information science yet another problem arose since not only Shannon but also Leon Brillouin, another founding father of cybernetics, related thermodynamic to informational entropy. Whereas Shannon was concerned with *potential* information and focused on the uncertainty present *before* a message is sent, Brillouin was concerned with *actual* information as the knowledge already available or what he called negative entropy or negentropy, and he focused on the uncertainty left *after* a message had been received. Consequently, for Shannon information and entropy were directly, for Brillouin inversely proportional. Mathematically, this is no problem, but for culture critics who employ 'entropy' not as a numerical but as an evaluative category, the choice of the one or the other definition without understanding their implications results in diametrically opposed judgments and explains why in literary criticism 'entropy' is used as both a positive (Shannon) and a negative concept (Brillouin). Moreover, whereas in physics and cybernetics the notions of 'order' and 'disorder' are value-free descriptions of molecular or informational aggregates, they inadvertently assume aesthetic implications which are all too often quite unwarranted when they are taken over by culture critics (see Arnheim).

It is hardly surprising that the entropy conundrum was picked up by science-fiction writers from H. G. Wells' *The Time Machine* to the 'ice-nine' in Kurt Vonnegut's *Cat's Cradle* and in many other tales in which it is often mixed up with the older notion of the Biblical apocalypse and consequently misunderstood as leading towards a death by instead of one of heat. But it seems almost impossible to relate the concepts of irreversible molecular dissipation and unavoidable informational entropy to what 'realistic' fiction deals with, namely individual human being and their everyday lives. The difficulties lie not only in the forbidding abstractness of a notion that cannot be visualized, but in the danger, which Henry Adams fell prey to, of applying concepts that relate only to closed systems to human beings who are open systems or, in the words of Ludwig von Bertalanffy, one of the founders of General Systems Theory, in solving "the violent contradiction between Lord Kelvin's degradation and Darwin's evolution, between the law of dissipation in physics and the law of evolution in biology" (39 f.).

*

It was a cocky twenty-one-year-old Cornell undergraduate named Thomas Pynchon who dared to face this challenge. In 1958/59 he wrote a story titled "Entropy" for his university writing class, rejected the advice to give it a less deterrent title, and had it published in 1960 in the prestigious *Kenyon Review*. A quarter of a century later, the by then famous novelist included this story in his collection *Slow Learner* (1984) and remembered:

> I happened to read Norbert Wiener's *The Human Use of Human Beings* (a rewrite for the interested layman of his more technical *Cybernetics*) at about the same time as *The Education of Henry Adams*, and the "theme" of the story is mostly derivative of what these two men had to say. A pose I found congenial in those days—fairly common, I hope, among pre-adults—was that of somber glee at any idea of mass destruction or decline . . . . Given my undergraduate mood, Adams's sense of power out of control, coupled with Wiener's spectacle of universal heat-death and mathematical stillness, seemed just the ticket. But the distance and grandiosity of this led me to short-change the humans in the story. I think they come off as synthetic, insufficiently alive. (xxiif.)

By now critics agree that Pynchon is "almost certainly more responsible than any others for leading American literary critics into an engagement with contemporary scientific ideas" (Porush 214), and his apprentice story is taken to be the first transliteration of the entropy notion in both its thermodynamic and its informational meaning into a fictional plot and as holding a crucial place in the ongoing realignment of what C. P. Snow diagnosed as "The Two Cultures" (see Freese).

The story's surface action alternates between two flats in a Washington, D.C. apartment-house "in early February of '57" (66). "Downstairs" (65) Meatball Mulligan's chaotic lease-breaking party is "moving into its fortieth hour" (65). Its inert guests have reached advanced stages of drunkenness, and a quartet of shop-talking jazz musicians is listening to records. Upstairs the reticent intellectual Callisto and his Eurasian girl friend Aubade live in a carefully arranged "hothouse jungle" (68), which has taken Callisto "seven years to weave together" (68) and in which he is presently dictating his memoirs and trying to prevent a sick bird from dying by transferring his body-heat to it. The tightly structured story, which is permeated by numerous musical analogies, is built like a fugue (see Redfield et al.; Pérez-

Llantada Auría), in which theme and counter-theme regularly alternate and "arabesques of order compet[e] fugally with the improvised discords of the party downstairs" (79).

Meatball's party is characterized by disorder, randomness and white noise, and in thermodynamic terms his apartment is an open system that is constantly replenished by new energy, since first "three coeds from George Washington" (71) join the party and later "five enlisted personnel of the U.S. Navy" (78), storm the flat which they mistake for a "hoorhouse" (78). By contrast, Callisto's "Rousseau-like fantasy" (68) world is characterized by minutely executed order, and in thermodynamic terms he has made it a "hermetically sealed" (68) closed system meant to function as "a tiny enclave of regularity in the city's chaos, alien to the vagaries of the weather, of national politics, of any civil disorder" (68). Callisto and Aubade never leave this enclave, and nobody else is ever allowed to enter it.

Since Pynchon has read Wiener, he is not content with constructing a fugal contrast between an open and a closed thermodynamic system. Therefore he has Saul, yet another tenant of the house, climb into Mulligan's apartment from the fire escape and tell him that he has just had a fight with his wife Miriam over, of all things, "communication theory" (75). He complains that they could not relate to each other because there is "Ambiguity. Redundance. Irrelevance, even. Leakage. All this is noise. Noise screws up your signal, makes for disorganization in the circuit" (76), and that their fight has ended with his wife throwing "a *Handbook of Chemistry and Physics*" (75) at him. Thus, Saul brings the notion of informational entropy, of noise, into the story. When he has complained about the unreliability of everyday language, Mulligan consoles him by saying "Well now, Saul, you're sort of, I don't know, expecting a lot from people. I mean, you know. What it is is, most of the things we say, I guess, are mostly noise" (77), and thus confirms Saul's complaint with a string of meaningless phrases that convey no information but only provide well-meant social noise.

Towards the end of the story, Mulligan's party erupts into several fights, and the tired host considers whether he should "lock himself in the closet" (84), which would function as a kind of miniature variation of Callisto's enclave, and just wait for his guests to leave, or whether he should "try to calm everybody down" (84) and thus "keep his lease-breaking party from deteriorating into total chaos" (84). He opts for the latter course, and while he begins to restore order, upstairs a despairing Callisto concludes that, with the sick bird having died in his hands and the outside temperature having remained the same for three days, maximum entropy has finally been reached. In a closing gesture of defiance, Aubade smashes the window and waits for "the moment of equilibrium" (85) between outside and inside temperature, at which all life will "resolve into a tonic of darkness and the final absence of all motion" (86).

This summary shows that Pynchon not only makes the entropy notion his controlling metaphor, but also manages to translate the different meanings of the concept into tellingly contrasted actions. Like F. Scott Fitzgerald, whose 'Valley of Ashes' in *The Great Gatsby* can be read as an early evocation of an entropic wasteland (see Scott), Callisto, the jaded intellectual, has studied at Princeton and traveled in Europe. He now lives in a home-made fantasy world with perfect "ecological balance" (68) and is a parodistic successor of that earlier American expatriate who had also dictated his despairing memoirs in Washington, and who had also done so in the third person. "Henry Adams, three generations before his own, had stared aghast at Power; Callisto found himself now in much the same state over Thermodynamics, the inner life of that power" (69). Like the misguided Adams of "A Letter to American Teachers of History" and "The Rule of Phase Applied to History," Callisto is afraid of "an eventual heat-death for the universe (something like Limbo: form and motion abolished, heat-energy identical at every point in it)" (69), and this is why he has retired into his artificial "sanctuary" (68), which he envisions, using Norbert Wiener's famous term, as "an enclave of regularity in the city's chaos" (68), that is, as an island of order in an ocean of irreversibly growing inertia, chaos and entropy. His *raison d'être* for his retirement from the increasingly disordered world he dictates to Aubade as follows:

> He had known all along, of course, that nothing but a theoretical engine or system ever runs at 100 percent efficiency; and about the theorem of Clausius, which states that the entropy of an isolated system always continually increases. It was not, however, until Gibbs and Boltzmann brought to this principle the methods of statistical mechanics that the horrible significance of it all dawned on him: only then did he realize that the isolated system—galaxy, engine, human being, culture, whatever—must evolve spontaneously toward the Condition of the More Probable. (72 f.)
>
> . . . .
>
> . . . he found in entropy or the measure of disorganization for a closed system an adequate metaphor to apply to certain phenomena in his own world. He saw, for example, the younger generation responding to Madison Avenue with the same spleen his own had once reserved for Wall Street: and in American 'consumerism' discovered a similar tendency from the least to the most probable, from differentiation to sameness, from ordered individuality to a kind of chaos. He found himself, in short, restating Gibbs' prediction in social terms, and envisioned a heat-death for his culture in which ideas, like heat-energy, would no longer be transferred, since each point in it would ultimately have the same quantity of energy; and intellectual motion would, accordingly, cease. (74)

In these musings Callisto sketches the history of the Second Law of Thermodynamics from Clausius to Boltzmann and Gibbs, and repeats Henry Adams' erroneous assumption that the thermodynamic notion of the heat-death provides "an adequate metaphor" for cultural developments and that the growth of entropy can be transferred from closed to open systems, from 'engines' to 'human beings' and 'cultures.' Callisto, whose ironic Greek name means 'the most beautiful,' has translated his despairing attitude into reality by imprisoning himself and Aubade, whose name refers to the troubadour's dawn song, in what he considers a hermetically sealed system. His artificial enclave of order is contrasted with the raucous party of his fellow tenant, whose name 'Meatball' makes him anything but a reticent intellectual and whose open house stands for life as constantly replenished by interaction with the environment. However, all of Meatball's guests have reached a state of stupor, so to the extent that their available energy has changed from free to bound, the party can also be read as an example of growing entropy.

The story ends with Aubade's defiant destruction of Callisto's enclave of order by smashing the window and with her and Callisto tensely waiting for the merging of their lives "into a tonic of darkness and the final absence of all motion" (86), and it opens with an epigraph from Henry Miller's *Tropic of Cancer*:

> Boris has just given me a summary of his views. He is a weather prophet. The weather will continue bad, he says. There will be more calamities, more death, more despair. Not the slightest indication of a change anywhere . . . . We must get into step, a lockstep toward the prison of death. There is no escape. The weather will not change. (65)

Most critics have identified Pynchon's position with that of Callisto, have read Aubade's gesture as "a dramatic suicide scene . . . rendering certain the entropic demise of the enclave," and have assumed that Mulligan's party group will "be overwhelmed by the same entropic cataclysm" (Redfield et al. 54). But such a reading is hardly borne out by the text, and "we should not jump to the conclusion that Pynchon is endorsing the metaphor [of entropy]" (Seed 137). In retrospect, Pynchon disparagingly called his apprentice story "a fine example of a procedural error beginning writers are always cautioned against. It is simply wrong to begin with a theme, symbol, or other abstract unifying agent, and then try to force characters and events to conform to it" (xxi). He critically remarked that he had "set things up in terms of temperature and not energy" and scornfully added: "I chose 37 degrees Fahrenheit for an equilibrium point because 37 degrees Celsius is the temperature of the human body. Cute, huh?" (xxiii). And he pointed out that he "put in the phrase *grippe espagnole*" because he was trying for "a sort of world-weary Middle-European effect" for Callisto, and said: "I must have thought this was some kind of post World War I spiritual malaise or something. Come to find out it means what it says, Spanish influenza, and the reference I lifted was really to the world-wide flu epidemic that followed the war" (xxvi).

In spite of the fact that "Entropy," which is cluttered with numerous open and hidden literary and cultural allusions ranging from William Faulkner and the Marquis de Sade to Djuna Barnes (see Bischoff; Coward; Hays; Hays et al.; Simons; Smetak), is rather too schematic and contains the weaknesses typical of all apprentice texts, it can be assumed that Pynchon, having read Adams and Wiener, knew enough about the entropy concept to realize that Callisto's attempt is as impossible as Adams' analogies were faulty. And this knowledge is expressed quite openly in the text. Not only is Mulligan's party a thermodynamically open system, which is constantly infused with new energy from outside, and not, as Tanner erroneously suggests, "a relatively closed system of people" (153). Callisto's enclave of order is also by necessity an open system, not only because it is insufficiently insulated against the noise drifting up from Meatball's party, but because it can only be sustained by being nourished from outside: "What they [Callisto and Aubade] needed from outside was delivered" (68). Consequently, Callisto's attempt at outwitting the Second Law is revealed as an intellectual self-deception, and this is underlined by the fact that the epigraph turns out to be a red herring. The Henry Miller quotation, whose reference to the unchanging weather serves as a metonymic anticipation of the imminent state of maximum entropy, turns out to be ironically misleading since it states only the opinion of Boris, the "weather prophet," but is immediately followed by the narrator's statement that "I am the happiest man alive" (1).

Charting the admissibility of analogies between the growth of thermodynamic entropy and the increase of disorder in cultural systems, Wiener had insisted that one needs "to keep these cosmic physical values well separated from any human system of valuation" (22), stressed that the Second Law pertains only to isolated systems, and maintained the possibility of "regions in which the entropy, defined according to a suitable definition, may well be seen to decrease" (23). His observation that "whether to interpret the second law of thermodynamics pessimistically or without any gloomy connotation depends on the importance we give to the universe at large, on the one hand, and to the islands of locally decreasing entropy which we find in it, on the other" (25), was certainly not lost on Pynchon. And this is why he depicts Wiener's choices by contrasting the otherworldly Callisto's fatalistic position with down-to-earth Mulligan's hopeful activities.

When the latter is faced with the alternative of either treating his party as if it were a steam-engine running down, that is, of going into hiding and letting the party peter out, or of employing his remaining energy in an attempt to restore some order and call "a repairman for the refrigerator, which someone had discovered was on the blink" (84), he decides to do the latter and thus proves that Callisto's fatalism is not the only option. The story, then, contrasts one man's pragmatic attempt at doing what he can in a given situation with another man's passive theorizing of cosmic doom, and there is no convincing reason why we should precipitately assume that Pynchon shares Callisto's view, whose endeavor he so clearly describes as futile.

The same, of course, is true with regard to informational entropy. Admittedly, Saul's fight with his wife and his talk with Meatball both describe and embody the ever-present danger that the ambiguity and redundancy of language might replace information by entropy and reduce communication to mere noise. But the very existence of Pynchon's story, which offers the possibility of a highly negentropic communication with its readers, shows that such pitfalls might be overcome. Thus, instead of sharing Callisto's deluded view of an imminent heat-death and of pessimistically depicting entropy as the inevitable "*finis* to the sum and total of being" (Steiner 83), Pynchon achieves the very opposite by relativizing the traditional understanding of the Second Law of Thermodynamics and by upholding the possibility that its relentless workings can at least be temporarily stalled through pragmatic activity and through the most negentropic action of all, the ordering work of the imaginative writer.

Pynchon's "Entropy" provides not only an important entry into his later novels about the mysterious workings of irreversibly growing entropy, but it also shows that the Second Law, which according to Eddington "holds . . . the supreme position among the laws of Nature" (74), can be made accessible to lay readers if it is suitably fictionalized. However, for a re-unification of Snow's Two Cultures one needs not only writers who can transliterate Boltzmann's equations into Meatball's Party and readers willing to make the effort to understand such translations, but also critics who are familiar with both worlds. In 1965, a disgruntled Kurt Vonnegut who suffered from being (dis)qualified as a science-fiction writer, observed:

> The feeling persists that no one can simultaneously be a respectable writer and understand how a refrigerator works, just as no gentleman wears a brown suit in the city. Colleges may be to blame. English majors are encouraged, I know, to hate chemistry and physics, and to be proud because they are not dull and creepy and humorless and war-oriented like the engineers across the quad. And our most impressive critics have commonly been such English majors, and they are squeamish about technology to this very day. (1 f.)

One would like to think that this has changed, but as late as 1999, Dietrich Schwanitz, a professor of English from Hamburg, published his bestseller *Bildung: Alles, was man wissen muss*, which promised that a reader who had mastered his hefty tome would be well "educated." In his survey of Western culture he maintained:

> Die Wendung von den zwei Kulturen wurde auch in Deutschland geläufig. Trotzdem hat der Appell von C. P. Snow so gut wie gar nichts bewirkt. Die naturwissenschaftlichen Kenntnisse werden zwar in der Schule gelehrt; sie tragen auch einiges zum Verständnis der Natur, aber wenig zum Verständnis der Kultur bei. Deshalb gilt man nach wie vor als unmöglich, wenn man nicht weiß, wer Rembrandt war. Wenn man aber keinen Schimmer hat, worum es im zweiten thermodynamischen Hauptsatz geht oder wie es um das Verhältnis der schwachen und starken Wechselwirkung des Elektromagnetismus und der Schwerkraft bestellt ist, oder was ein Quark ist, obwohl die Bezeichnung aus einem Roman von Joyce stammt, dann wird niemand daraus auf mangelnde Bildung schließen. So bedauerlich es manchem erscheinen mag: Naturwissenschaftliche Kenntnisse müssen zwar nicht versteckt werden, aber zur Bildung gehören sie nicht. (482)

In 2001, an annoyed professor of the history of science, Ernst Peter Fischer from Konstanz, answered Schwanitz with another bestseller entitled *Die andere Bildung: Was man von den Naturwissenschaften wissen sollte*. He angrily accused Schwanitz of hiding his scientific ignorance by declaring only that knowledge as educationally relevant which existed within his limited horizon, and he bitterly complained about the "Hochmut eines literarisch und philosophisch Gebildeten gegenüber den Leistungen der Naturwissenschaften" (10).

It seems, then, that despite the efforts of writers from Pynchon through Gaddis to Powers to reunite Snow's "two cultures" and despite many critical attempts to further such a re-unification (see Freese et al. 2004; 2004a) the traditional division of labor between nature and culture still exists. With spokespeople of the two camps lustily accusing each other of one-sidedness, and with the ongoing debate about cloning illustrating the mutual incomprehension between worried humanists and progressive scientists, the present German situation seems to be an exact replica of the scene about which Snow famously complained over forty years ago. Apparently, Joseph Bronowski's plea for common ground has remained widely unheard: "Science and the arts today are not as discordant as many people think. It is the business of each of us to try to remake that one universal language which alone can unite art and science, and layman and scientist, in a common understanding" (13). Our academies are still divided into schools for those who invent and build the machines that dominate our life and those who are expected to ponder the social repercussions and moral implications of these machines. And there are even some who argue that the division should be upheld, as is illustrated by a 1997 review of Ian McEwan's novel *Enduring Love*, in which an enraged British critic called

> for an immediate worldwide moratorium on novelists reading works of science. Like oceans plundered of whales, science-books have become overfished by voracious, imaginative writers. You can't pick up a novel these days without being bombarded by Heisenberg's Uncertainty Principle, or the latest theories on Darwinism. Popular science now occupies ample shelf-room in every bookshop and a prominent place in bestseller lists. Novelists should tell us stories, not recite particle physics. I'm all in favour of the novel of ideas, but at least let the ideas be the author's own. An author's individuality is drowned in this sea of science. (Connolly 34)

Thus, in spite of some promising progress, much still needs to be done to reconcile science and the humanities.

Works Cited

Adams, Henry. *The Degradation of the Democratic Dogma*. New York: Capricorn Books, 1947.

Arnheim, Rudolf. *Entropy and Art: An Essay on Disorder and Order*. Berkeley, Los Angeles and London: University of California Press, 1971.

Auerbach, Felix. *Ektropismus oder die physikalische Theorie des Lebens*. Leipzig: Wilhelm Engelmann, 1910.

---. *Die Weltherrin und ihr Schatten: Ein Vortrag über Energie und Entropie*. Jena: Gustav Fischer, 1902.

Bergson, Henri. *L'Evolution créatrice* (1907). In his *Œuvres*. Paris: Presses Universitaires de France, 1970.

Bischoff, Peter. "Thomas Pynchon, 'Entropy' (1960)." *Die amerikanische Short Story der Gegenwart: Interpretationen*. Ed. Peter Freese. Berlin: Erich Schmidt, 1976. 226-236.

Brillouin Leon. *Science and Information Theory*. New York: Academic Press, 2nd ed., 1962.

Bronowski, J. *The Common Sense of Science*. Cambridge: Harvard University Press, 1953.

Clausius Rudolf. "Über verschiedene für die Anwendung bequeme Formen der Hauptgleichungen der mechanischen Wärmetheorie." *Annalen der Physik und Chemie* 125 (1865): 353-400.

Connolly, Cressida. "Cleverly Done." *Literary Review* September 1997: 34.

Cowart, David. "Science and the Arts in Pynchon's 'Entropy.'" *College Language Association Journal* 24 (1980/81): 108-115.

Eddington, Sir Arthur. *The Nature of the Physical World*. Ann Arbor: University of Michigan Press, 2nd ed., 1963.

Eliot, T. S. *The Complete Poems and Plays of T. S. Eliot*. London: Faber and Faber, rpt. 1981.

Engels, Friedrich. *Dialektik der Natur*. Berlin: Dietz Verlag, 1973.

Fischer, Ernst Peter. *Die andere Bildung: Was man von den Naturwissenschaften wissen sollte*. München: Ullstein, 2001.

Freese, Peter. *From Apocalypse to Entropy and Beyond: The Second Law of Thermodynamics in Post-War American Fiction*. Essen: Die Blaue Eule, 1997.

Freese, Peter and Charles B. Harris (eds.). Science, *Technology, and the Humanities in Recent American Fiction*. Essen: Die Blaue Eule, 2004.

---. *The Holodeck in the Garden: Science and Technology in Contemporary American Fiction*. Normal, Il.: Dalkey Archive Press, 2004a.

Hays, Peter L. "Pynchon's 'Entropy': A Russian Connection." *Pynchon Notes* 16 (1985): 78-82.

---. and Robert Redfield. "Pynchon's Spanish Source for 'Entropy.'" *Studies in Short Fiction* 16 (1979): 327-334.

Joule, James Prescott. "On Matter, Living Force, and Heat." *The Scientific Papers*. London: Dawson of Pall Mall, 1963 [reprint of the 1st ed. of 1887). Vol. I. 265-276.

Kannegiesser, Karl-Heinz. "Zum zweiten Hauptsatz der Thermodynamik." *Deutsche Zeitschrift für Philosophie* 9 (1961): 841-859.

Maxwell, James Clerk. *Theory of Heat*. Westport, CT: Greenwood Press, 1970. Reprint of the 3rd ed. of 1872.

Miller, Henry. *Tropic of Cancer*. New York: Grove Press, 1961.

Nietzsche, Friedrich. *Der Wille zur Macht: Versuch einer Umwertung aller Werte*. In *Werke*, Kritische Gesamtausgabe. Eds. Giorgio Colli and Mazzino Montinari. Berlin and New York: de Gruyter, 1967 ff.

Pérez-Llantada Auría, Carmen. "Beyond Linguistic Barriers: The Musical Fugue Structure of Thomas Pynchon's 'Entropy.'" *Cuadernos de Investigacíon Filologica* 17.1-2 (1991): 127-140.

Planck Max. *Eight Lectures on Theoretical Physics Delivered at Columbia University in 1909*. New York: Columbia University Press, 1915.

Porush. David. "'Unfurrowing the Mind's Plowshare': Fiction in a Cybernetic Age." *American Literature and Science*. Ed. Robert J. Scholnick. Lexington: University Press of Kentucky, 1992. 209-228.

Pynchon, Thomas. *Slow Learner*. Toronto and New York: Bantam Books, 1985.

---. *The Crying of Lot 49*. New York: Bantam Books, 1967.

Redfield, Robert, and Peter L. Hays. "Fugue as a Structure in Pynchon's 'Entropy.'" *Pacific Coast Philology* 12 (October 1977): 50-55.

Schwanitz, Dietrich. *Bildung: Alles, was man wissen muß*. Frankfurt a.M.: Eichborn, 1999.

Scott, Robert Ian. "A Sense of Loss: Entropy vs. Ecology in *The Great Gatsby*." *Queen's Quarterly* 82 (1975): 559-571.

Seed, David. "Order in Thomas Pynchon's 'Entropy.'" *Journal of Narrative Technique* 11 (1981): 135-153.

Shannon, Claude and Warren Weaver. *The Mathematical Theory of Communication*. Urbana, Chicago and London: University of Illinois Press, rpt. 1972.

Simons, John. "Third Story Man: Biblical Irony in Thomas Pynchon's 'Entropy.'" *Studies in Short Fiction* 14 (1977): 88-93.

Smetak, Jacqueline R. "Thomas Pynchon's Short Stories and Jung's Concept of the Anima," *Journal of Evolutionary Psychology* 11, 1-2 (March 1990): 178-194.

Snow, C. P. *The Two Cultures. And a Second Look*. Cambridge: Cambridge University Press, 1965.

Steiner, George. *Proofs and Three Parables*. London: Faber and Faber, 1992.

Tanner, Tony. *City of Words: American Fiction 1950 – 1970*. London: Jonathan Cape, 1971.

Thomson, William, Lord Kelvin. "The Sorting Demon of Maxwell" (1879). Rpt. in his *Popular Lectures and Addresses*. 3 vols. London and New York: Macmillan, 1889. Vol. I, *Constitution of Matter*. 137-141.

---. "On a Universal Tendency in Nature to the Dissipation of Universal Energy." *Philosophical Magazine* 4 (1852): 304-306.

Tribus, Myron. "Thirty Years of Information Theory." *The Maximum Entropy Formalism: A Conference Held at the Massachusetts Institute of Technology on May 2-4, 1978*. Eds. Raphael D. Levine and Myron Tribus. Cambridge, Mass.: MIT Press, 1979. 1-14.

von Bertalanffy, Ludwig. *General Systems Theory: Foundations, Development, Applications*. London: Allen Lane the Penguin Press, 1971.

von Foerster, Heinz. "Disorder/Order: Discovery or Invention?" *Disorder and Order: Proceedings of the Stanford International Symposium (Sept. 14-16, 1981)*. Ed. Paisley Livingston. Saratoga: Anma Libri, 1984. 177-189.

von Helmholtz, Hermann. "Über die Wechselwirkung der Naturkräfte und die darauf bezüglichen neuesten Ermittelungen der Physik." Ed. Hermann von Helmholtz. *Populäre wissenschaftliche Vorträge*. 2nd ed.; 2. Heft. Braunschweig: Friedrich Vieweg und Sohn, 1876. 101-133.

Vonnegut, Kurt Jr. *Wampeters, Foma, & Granfalloons*. New York: Delacorte, 1974.

Wiener, Norbert. *The Human Use of Human Beings: Cybernetics and Society*. Boston: Houghton Mifflin, 1950.

Technoscience and its Publics: Theories and Practices

Bio Science: Genetic Genealogy Testing and the Pursuit of African Ancestry

Alondra Nelson

> *[O]ur biographies are written, at least in part, in terms of structural chemistry.*
>
> *Margaret Lock*[1]

Recent years have seen the introduction of commercial genetic testing for genealogical purposes. Media accounts of consumers' experiences with said analysis often suggest that genetic genealogy test results irrevocably transform how test-takers understand their selves, their communities and their families; in other words, that individual and collective identities are "geneticized." As I describe in this essay, however, test-takers' responses to genetic genealogy reflect a negotiation between known or desired biographical information and molecular biology, between bios and *bios*. Drawing upon ethnographic fieldwork and interviews, I argue that although there is some acquiescence to genetic thinking about ancestry, and by implication, race and ethnicity, among African American and black British consumers of genetic genealogy testing, test-takers also adjudicate between varied sources of genealogical information and from these construct meaningful biographical narratives. By engaging in highly situated self-fashioning and interpreting genetic test results in the context of their "genealogical aspirations," consumers mediate and move beyond the "two cultures."

The decoding of the human genome precipitated a change of paradigms in genetics research, from an emphasis on what then President Bill Clinton, in his announcement of this scientific achievement, described as "our common humanity" (White House F8) to a concern with molecular-level differences among individuals and groups. This shift in research focus from lumping to splitting spurred ongoing disagreements among scholars in the social and biological sciences about whether genetic markers can and should be used to distinguish human groups. One fulcrum on which this debate has hinged is the question of the epistemological status of "race."

These divergent perspectives on the definition and meaning of "race" can be generally characterized as pragmatism and "naturalism."[2] I define race pragmatists

1 Here Lock is paraphrasing chemist H.E. Armstrong.

as those scholars who emphasize the practical outcomes for lived experience of the historically-contingent processes of racialization. For them, "race" is not a biological fact, but a social invention, better understood as an index of power—one that structures access to resources including healthcare, education, and housing—than as a register of inherent human difference.[3] The pragmatist position was succinctly articulated in a 2001 editorial in the *New England Journal of Medicine* that stated that "[r]ace is a social construct, not a scientific classification" (Schwartz 1392). Pragmatists bolster their arguments by citing human population geneticists' findings that humans are 99.9% alike and that intra-group genetic differences vary far more than inter-group ones.[4]

In contrast, the contemporary race naturalist position can be summarized with the following assertion: "[n]ature makes differences between individuals. These differences are real, not constructed" (Hacking 103).[5] Extending this logic, contemporary race naturalists contend that humans can be classified into groupings that confirm the biological reality of "race."[6] These claims are based on novel techniques that allow scientists to locate genetic variants shared *across* human groups, but differently distributed *among* them, and to subsequently ascribe racial and ethnic distinctions to this statistical spectrum. Two recent, influential papers published in *Science* and the *American Journal of Human Genetics,* for example, argue that genetic markers can be used to predict an individual's geographic origins, with the concepts of "population," "origin," and "geography" serving as a proxy for "race."[7]

2 For more on naturalism, see Ian Hacking. These concepts are, of course, generalizations of a range of perspectives about the meaning and significance of race. Nonetheless, I think they serve as a useful analytic for describing debates about race and genetics—which have not abided disciplinary boundaries—without reducing them to a contest between natural scientists and social scientists, and thus, for showing the imbrication of naturalist and pragmatist positions. Both Donna Haraway and Jennifer Reardon, for example, have demonstrated that racial pragmatists include social and natural scientists who invoke findings from the natural sciences to support their argument that race is a biological fallacy.

3 Fuller citation of the pragmatist approach can be found in Michael Omi and Howard Winant; David Theo Goldberg; Audrey Smedley; Robert Schwartz; and American Sociological Association.

4 See Richard Lewontin and American Anthropological Association. Both Lewontin's and the American Anthropological Association's texts exemplify the pragmatist position on intra- and inter-group genetic variations.

5 "Contemporary" is used here to mark a difference between 18th and 19th century arguments about racial taxonomies and assertions about innate human difference, and current naturalist positions articulated by scholars who are typically aware of the history of scientific racism and who disavow it, even as they maintain their commitment to the reality of human types. Naturalists conceive of their position as distinct from the legacy of scientific racism, while pragmatists consider the naturalist position to be an extension of this conceptual trajectory.

6 Sarich and Miele's text explicitly illustrates this contemporary race naturalist position.

7 See M. J. Bamshad, et al; and N. A. Rosenberg, et al.

The stakes of the pragmatist-naturalist debate are high. Naturalists underscore the urgency of their position with the argument that pressing healthcare concerns oblige researchers to use racial and ethnic categories as "starting points" for genetic research (Burchard et al. 1174). Pragmatists conversely and aptly point to the intertwined history of science, medicine, and race that is punctuated with instances in which scientific theories of human difference are used to justify forms of discrimination from chattel slavery to segregation. They contend that the new splitting techniques in genetics threaten to sustain and compound this deplorable legacy.[8]

Yet, both naturalist and pragmatist conceptions of "race" have drawbacks. Invoking scientific objectivity, naturalists may abjure responsibility for participation in research programs that presume inherent human difference and for how subsequent findings are socially reified (Cooper, Kaufman and Ward 1169). For example, Risch et al. attempt to dissociate genetic research into group differences from both its social origins and social effects when they write that "[t]he notion of superiority is not scientific, only political, and can only be used for political purposes" (11). Alternatively, while pragmatists attend to past injuries and potential risks of scientific racism when they consider recent developments in genetics research, some do not fully appreciate how "race" can be a non-deterministic biological "discourse about the body."[9] Condit's research on the public understanding of genetics, as outlined in *The Meanings of the Gene,* is instructive on this point. In a study of the connotations of the word "blueprint" as a descriptor of genes, Condit found that respondents used the word to evoke both schematic and schema, that is, an immutable infrastructure of human nature as well as a malleable constellation of constitutive facets of identity. This research shows that there is considerable variance in how the influence of genes is construed by the public, and suggests that the presence of such multiple interpretations bears consideration in contemporary investigations of race and ethnicity. Rejecting a reductive understanding of "race" as genetic fact should not preclude investigations into the ways in which biological discourses contribute to racial formation processes beyond the lab and the clinic. Indeed, "race" and ethnicity have never been the products of a single domain of knowledge or influence, such as of the biosciences or the social sciences exclusively.[10] Rather, their significance has always been constituted, simultaneously

[8] For more on the pragmatist critique of the dangers portended by the naturalist position, and its interpretation of these new techniques, see Lee Baker; Troy Duster (both publications).

[9] I borrow the phrase "discourse about the body" from Donna Haraway's 1997 text, *Modest_Witness.* Indeed, following Haraway's investigation of the constructed nature of bioscientific knowledge in *Simians, Cyborgs, and Women*, I seek to highlight how "bodies are made" in scientific and other practices.

[10] See Baker's *From Savage to Negro* on the history of anthropology; Gilroy's *Against Race* on "raciology" and humanism; Goldberg's *Anatomy of Racism* on the socio-historical contexts of racism; Jacobson's *Whiteness of a Different Color* on ethnicity and whiteness in the USA; Lo-

"socially" and "naturally," through assemblages of discourses, concepts, ideologies, and practices enacted at various social locations (Fausto-Sterling 2-4).[11] Accordingly, investigations of "race" and ethnicity in the genomics era need to be pursued from many perspectives and on many scales. In this paper, I endeavor to triangulate the naturalist-pragmatist "binary trap" (Ossorio and Duster 115) through an examination of the consumption[12] of genetic genealogy testing – the use of DNA analysis for the purpose of inferring ethnic or racial background and aiding with family history research.[13]

Knowledge derived from genetic science has increasingly been used to explain ever-growing aspects of the social world. The proliferation of genetic genealogy testing—DNA analysis which purveyors claim provides scientific substantiation of an individual's ancestral origins based on the comparison of his or her DNA against a database of genetic samples from a statistically constituted social group—appears to be one example of this increasing "geneticization."[14] The questions I pose are whether and how these technologies come to "geneticize" racial and ethnic identities. The decision to employ genetic testing for genealogical purposes could be viewed as a sign that test-takers are confident about the underlying assumptions of this form of genetic analysis as well as in naturalist conceptions of human difference. However, as I describe below, while the geneticization of race and ethnicity may be the basic logic of genetic genealogy testing, it is not necessarily its inexorable outcome (Rose 176-177). My research shows that, to the contrary, the scientific data supplied through genetic genealogy is not always accepted as definitive proof of identity; test results are valuable to "root-seekers" to the extent that they can be deployed in the construction of their individual and collective

pez's *White by Law* on citizenship law; Omi and Winant's *Racial Formation* on racial formation theory.

[11] See Omi and Winant's *Racial Formation* and Gilroy's *Against Race* for discussions of the "social" and "natural" dimensions of "race."

[12] The ethnographic data for this paper are drawn from my interactions with persons who have *purchased* genetic genealogy testing as well as individuals who participated in a study in which such testing was employed, but for which they *paid no fees*. Accordingly, the word 'consumption' here is meant to broadly connote "use." "Consumer" is used only in reference to people who have purchased test kits. When referring to both groups of informants, and to avoid confusion, I use the terms "test-taker" and "root-seeker."

[13] Here, I echo Ossorio and Duster in stressing the need for debates about race and genetics to move beyond currently entrenched perspectives, and encourage scholars to truly understand the risks and possibilities presented by new racial (in)formation processes. As they contend, this debate has produced more heat than light: it fails to illuminate "the complex interplay between biological and social aspects of human taxonomies" and the "social processes that can create biological feedbacks" (116). In this paper, I seek also to understand how ideas and techniques from genetic science in turn create social feedback that can be selectively incorporated into individual and group identity formation. Pilar Ossorio and Troy Duster 115.

[14] See Abby Lippman 1991 and 1998.

biographies. Root-seekers align *bios* (life) and bios (life narratives, life histories) in ways that are meaningful to them.[15] These users of genetic genealogy interpret and employ their test results in the context of personal experience and the historically shaped politics of identity.[16] They actively draw together and evaluate between many sources of genealogical information (genetic and otherwise) and from these weave their own ancestry narratives.

This paper draws upon interviews with and ethnographic fieldwork among persons of African descent who have made use of one or more categories of genetic genealogy testing. Since 2003, I have observed events and conferences at which genetic genealogy testing was discussed or offered including meetings at churches, libraries, and universities in England and the USA.[17] In Britain, my research has centered on the London metropolitan area where I attended a gathering of subjects in the research project and BBC documentary, "Motherland: A Genetic Journey"; I also interviewed several study participants and the program's producers. At sites throughout the USA, I attended gatherings of "conventional" genealogists – amateur researchers who employ archives and oral history, among other sources, to reconstruct family history. Because of their demonstrated interest in ancestry tracing, conventional genealogists are targeted by purveyors of genetic testing services. In this process, I also have become a genealogist and my own research subject. I am currently conducting research on my family's history that traverses the southern USA and Jamaica. I am a member of the Afro-American Historical and Genealogical Society (AAHGS) and of a local AAHGS chapter in New York city. I intend to purchase at least one genetic genealogy test after I have completed more family history research.

In addition to archival excavation, the practice of genealogy now involves considerable technical mediation: Family Tree Maker and other computer software programs that assist genealogists in constructing pedigree charts and rationalizing

[15] My thanks to Stefan Helmreich for suggesting this phrasing. Root-seekers alignment of *bios* and bios is similar to the process of "categorical alignment" described by Steven Epstein.

[16] See Margaret Lock; Keith Wailoo and Stephen Pemberton.

[17] In the larger project that produced this paper, in addition to following the experiences of test-takers, I also track other "reconcialtion projects," ways in which how genetic genealogy testing is put to the purpose of responding to a variety of social issues related to the legacy of racial slavery. Thus, in addition to investigating the experiences of test-takers in the USA and the UK, I explore the development of genetic genealogy testing for African Ancestry in the African Burial Ground research project, in which investigators examined human remains uncovered at a long-forgotten African cemetery in downtown Manhattan in 1991. I follow the use of these tests as possible evidence in an ongoing class action suit for reparations for slavery in the USA. I also am interested in how these tests are used for purposes of commemoration, such as in the case of the former Connecticut slave Venture Smith and his wife Margaret "Meg" Smith, who were exhumed in 2006 in the hopes of discerning their African ethnicity and completing the historical record.

the large amounts of information the activity requires; email list-servs dedicated to discussion of the technical aspects of genetic genealogy testing; and websites at which test-takers can compare DNA results in order to establish degrees of relation. Since contemporary genealogical research is a substantially technological pursuit, my research necessarily involved "virtual" ethnography.[18] Specifically, I observed and participated in a virtual community whose members share an interest in tracing African ancestry. Community discussions encompassed varied topics related to the practice of genealogy. One forum is dedicated to discussions of DNA testing. In virtual genealogical settings, members also dialogue about the science behind genetic ancestry tracing; display expertise through their command of jargon and recent genetics research or developed through prior experience with one or more testing companies; circulate topical scientific papers and newspaper articles; and share genetic genealogy test results and their feelings about them. As a participant observer of genealogical settings, my involvement in this online community primarily consisted of discussions with genealogists, both on the public listserv and "off channel" – that is, in private online conversations. There were "nodes" of overlap and continuity between the on-line and off-line communities I inhabited and, as I describe below, I came to know several members of this virtual community personally through interviews and at genealogical gatherings.[19]

In what follows, I discuss these African American and black British consumers of genetic genealogy testing whose accounts provide a window to this emerging practice of using genetic and socio-historical resources to constitute their identities and thereby also to constitute race and ethnicity in the age of genomics.[20] I trace the historical and cultural precedents of black root-seeking, and then discuss three categories of genetic genealogy testing and the information each provides. Turning to the experiences of test-takers, I consider whether and how genetic genealogy test results are incorporated into individual and collective biographies. Extending the concept of "objective self-fashioning" (Dumit 44) to "*affiliative* self-fashioning," I argue that genetic genealogists exercise some control over the interpretation of their test results, despite the presumption of their conclusiveness. Among other factors, test-takers' negotiation of test outcomes may be generated by a disjuncture between genetic and other types of evidence about ancestry that can elicit an affect I term "genealogical disorientation." Genetic genealogy testing may thus amplify possibilities for subject-formation and ancestral affiliation, rather than simply re-

[18] For details on virtual ethnography, see Christine M. Hine; Daniel Miller and Don Slater; Stefan Helmreich; and Deborah Heath et al.

[19] See Heath et al.

[20] I use the term "constitute" rather than "construct" because the former evokes the processes of racialization and ethnicization *as well as* debates about the constitutive elements or the "stuff" of which race and ethnicity is made (for example, gene variants, cultural practices, social contexts, power relations, and so on)

ducing them to genetic determinants. I conclude that, contrary to both naturalist and pragmatist arguments, genetic genealogy testing provides a locus at which racial and ethnicity are constituted at the nexus of genetic science, kinship aspirations, and strategic self-making.

Genetic Root-Seeking and the Usable Past

Until recently, for persons of African descent and others, pursuing one's family history has typically entailed genealogical excavation of the type depicted in Alex Haley's best-selling book *Roots: The Saga of an American Family* – a novelized account of Haley's efforts to trace his ancestral lineage back to the African continent. As is widely acknowledged, Haley's project—the book and the award-wining television mini-series adapted from it—prompted an international conversation on racial slavery and its consequences. Less recognized, but equally important, Haley's narrative established an expectation among a generation of readers and viewers in the USA and abroad that recovering ancestral roots was not only desirable, but also possible.[21]

Genealogists of African descent frequently reference *Roots* when describing how their interest in family history research was piqued. Elisabeth's (a pseudonym)[22] experience is typical of the genealogists with whom I spoke—typically aged 40 years or older, college educated, and predominantly female—who as teenagers or young adults were inspired by Haley's example.[23] I first encountered Elisabeth, a computer scientist in her late forties, in an online community of black genealogists to which we both belong and interviewed her at her home in the northwestern USA in 2004. To the rhythm of her placing and removing from the oven

[21] Genetic genealogy testing is employed by other diasporic social groups whose historical experiences of migration, dispersal, and persecution have made it difficult to document genealogical information including Irish (see Catherine Nash's "Genetic Kinship") and Jewish (see Nadia Abu El-Haj's *Facts on the Ground*) communities. These services are also widely used for religious reasons. For example, genealogy is an important part of the after-life cosmology of the Church of Jesus Christ of Latter Day Saints.

[22] My informants reveal sensitive and personal information about issues of identity, community, and belonging, and so I use pseudonyms in order to protect their privacy. I do, however, use the actual names of the three participants in the *Motherland* documentary who, by participating in the film, have chosen to make their accounts public, and of the documentary producers whose roles are also public knowledge. I also use the name of the leader of the Motherland Group with his permission. The names of all other members of that group are pseudonyms.

[23] The predominance of women among my informants and, in genealogical communities more generally, is consistent with the literature on "kinkeeping," the term used by Carolyn J. Rosenthal to describe the gendered work of maintaining family ties, through activities such fostering communication between members or providing emotional and financial aid to them. With genealogical practices, kinkeeping involves the work of connecting *past* and *present* kin.

the cookies she was baking for her pre-teen son, she described the chain of events that had led her to become a genealogist and, some decades later, a genetic genealogist. In particular, she waxed nostalgic about a presentation by Alex Haley at her Midwestern high school that had stimulated her interest in genealogy:

> Haley came to my high school in 1970. This was before *Roots* came out. He had a *Reader's Digest* article about it out and he was on the road just telling everyone about how he traced Kunte Kinte. And, I was in 9th grade and I just sat there mesmerized...Actually, I have a copy of the tape [of Haley's presentation]. I got in contact with my old high school civics teacher, out of the blue, last year. And, he says, 'You know, I was going through stuff and I found this old Alex Haley tape. I didn't know what to do with it – would you like it?' Of course! And it's phenomenal! . . . It was just a fascinating talk; it really was. That's when I got bit by the genealogy bug.

Elisabeth's friend, Marla, expressed similar sentiment about Haley's influence when I met with her. Although she made a start at genealogical research in the 1960s, following the death of the eldest member of her extended family, it was not until a decade later, when she attended a lecture by the author at a local community college, that her interest was galvanized. This encounter impressed upon her that a non-specialist researcher could employ insurance records, land deeds, slave-ship manifests, and family history libraries to trace her roots to Africa.[24] As she explained to me,

> it was interesting to hear him talk about . . . going to the Mormon temple and going to Lloyd's of London and all of that. I never figured that I would have access to those kinds of records . . . I never ever thought that the average person could have accessed it. So, I never anticipated being able to . . . go back to slavery.

Across the Atlantic Ocean, the "Roots" mini-series elicited feelings of "deep shame and embarrassment" in a teen-aged Beaula McCalla who, as an adult, endeavored to cope with these emotions by organizing African-centered programming in Bristol, England, and, eventually, by participating in a television documentary entitled *Motherland: A Genetic Journey*.[25]

In notable contrast to the ubiquity of what might be called the "*Roots* moment" among root-seekers, very few of my informants have been able to complete their

[24] Marla also felt a special connection to Haley's mission because his father had been a professor at her college.

[25] Beaula McCalla, Personal interview, 2003.

familial lineages using standard genealogical methods like those Haley used.[26] Persons wishing to trace their African ancestry beyond the 19th century face many hurdles to the achievement of this goal—genealogists speak of coming up against a "brick wall"—principally, the scarcity of written records from the era of the slave trade. Consequently, DNA analysis appeals especially to root-seekers whose prior efforts have failed to yield information sufficient for extensive genealogical reconstruction.

Genetic genealogy testing emerged from techniques developed in molecular genetics, human population genetics, and biological anthropology.[27] Three principal tests are offered by the growing number of companies that sell DNA analysis for genealogical purposes. Rather than taking up the companies' technical or brand descriptions, I categorize the tests according to the type of information each imparts because, as I discuss below, the forms of social orientation that the test outcomes suggest are of primary importance to root-seekers. My informants purchased particular genetic tests in order to fulfill distinct "genealogical aspirations," such as corroboration of a multicultural background or assignment to an ethnic community. With this in mind, I classify these tests as ethnic lineage, racio-ethnic composite, and spatio-temporal.

Ethnic lineage testing draws on the unique features of *Y-chromosome* DNA (Y-DNA) and *mitochondrial* DNA (mtDNA) to infer ancestral links to contemporary nation-states or cultural groups. Y-DNA is passed virtually unchanged from fathers to sons and can be used to trace a direct line of male ancestors; mtDNA, the energy catalyst of cells, is inherited by male and female children exclusively from their mothers, and contains "hypervariable" segments that are conducive to comparison and thus useful for uncovering matrilineage. Using both forms of ethnic lineage testing, a consumer's DNA is searched against a testing company's reference database of genetic samples. If the sample and the reference DNA match at a set number of genetic markers (typically eight or more), an individual can be said to have shared a distant maternal or paternal ancestor with the person who was the source of the matching sample in the reference population. Several companies offer this type of testing, including African Ancestry (africanancestry.com) and Family Tree DNA (ftdna.com). A typical ethnic lineage result may inform a test-taker that her mtDNA traced to the Mende people of contemporary southern Sierra Leone.

With spatio-temporal testing, a consumer's DNA sample is classified into a haplogroup (sets of single nucleotide polymorphisms [SNPs] or gene sequence

[26] Haley's research is controversial. He was charged with plagiarizing from the novel *The African*—a matter that he settled out of court—and was also accused of fictionalizing much of his account of his ancestors' lives.

[27] See Luigi Cavalli-Sforza, Paolo Menozzi and Alberto Piazza; Michael F. Hammer; Mark A. Jobling and Chris Tyler-Smith; and Skorecki et al.

variants that are inherited together) from which ancestral and geographical origins at some point in the distant past can be inferred. This form of analysis was made possible by the ambitious Y-DNA and mtDNA mapping research that resulted in theories about the times and places at which various human populations arose.[28] Family Tree DNA supplies customers with haplogroup information as does National Geographic's Genographic Project. Based on a match with the mtDNA-derived L1 haplogroup, a customer employing this test can receive a result indicating that her ancestors lived in Africa approximately 100,000 years ago.

Racio-ethnic composite testing involves the study of nuclear DNA—which is unique to each person (identical twins excepted, although this is now being debated) and consists of the full complement of genetic information inherited from parents—for the purpose of making claims about one's ancestry. A DNA sample is compared with panels of proprietary SNPs that are deemed to be "informative" of ancestry. Algorithms and computational mathematics are used to analyze the samples and infer the individual's "admixture" of three of four statistically-constituted categories—African, Native American, East Asian, and European—according to the presence and frequency of specific genetic markers said to be predominate among, but importantly, not distinctive to, each of the "original" populations.[29] This form of analysis was developed and is principally offered by the Ancestry by DNA division of DNAPrint Genomics as well as by other companies that use its techniques, such as the Genetic Testing Laboratories in New Mexico and UK-based International Biosciences. A hypothetical customer might learn his composite to be 80% African, 12% European and 8% Native American.

Each of these tests thus offers a different window into the past, and root-seekers demonstrate different interests and preferences based on their genealogical aspirations. Racio-ethnic composite testing has proved unsatisfactory to some root-seekers who want to re-create Alex Haley's *Roots* journey in their own lives. Although composite testing analyzes an individual's full genome (rather than a section of it, as is the case with ethnic lineage and spatio-temporal testing), its results nevertheless lack specificity and usefulness for some users.[30] Cecily was one of

[28] See R. Cann, M. Stoneking and A. Wilson; also Luigi Cavalli-Sforza, Paolo Menozzi and Alberto Piazza.

[29] For a detailed discussion of racio-ethnic composite or 'AIMS' analysis, see Duana Fullwiley, "The Biologistical Construction of Race."

[30] To be sure, the usefulness of test results depends on the test-taker and the particular questions he or she seeks to answer through genetic genealogy testing. Here, I am highlighting how test preferences are shaped by the problems to which they are applied. It should be also be noted, however, that many of the test-takers I encountered used more than one type of genetic genealogy analysis, typically to compare results received from different companies or to obtain new information from a company from which services were purchased previously (for example, when a company releases a more robust form of test that employs more markers or has added a significantly larger number of samples to its database)

them. We met at the 2005 annual meeting of the AAHGS. As we sat near the display booth of the African Ancestry company from which she had previously purchased an ethnic lineage test, I asked whether she planned also to pursue racioethnic composite testing. In response, she declared, "I don't need to take that test. We're all mixed up. We know that already." Somewhat similarly, spatio-temporal testing results may be deemed too remote by some root-seekers. Marla, who in addition to the genealogy chapter she leads with Elisabeth also moderates an internet forum dedicated to discussion of DNA testing for genealogical purposes, has purchased several tests. An mtDNA test purchased from African Ancestry matched her with the Tikar people of Cameroon. As I have found is frequently the case, Marla's initial testing experience stimulated further curiosity about her ancestry, rather than satisfying it fully. She then purchased a composite test for herself and also paid for three family members to have ethnic lineage testing from Trace Genetics (a testing company known for its large database of Native American reference samples, which was purchased by DNAPrint Genomics, in 2006). For a fourth round of testing, Marla sought to find out more about the maternal line of her deceased father. As a seasoned genealogist, she knew that this information could be accessed if she had a paternal second cousin's DNA analyzed. In an email exchange between Marla and I that followed from a conversation at her home, she detailed Family Tree DNA's spatio-temporal analysis of her cousin's genetic sample:

> The mtDNA of my 1st cousin's daughter (paternal grandmother's line) traced to 'Ethiopia' and ±50,000 years ago. It is Haplogroup L3 which [according to the information provided by the company] 'is widespread throughout Africa and may be more than 50,000 years old'. Her [the cousin's] particular sequence 'is widespread throughout Africa' and has its 'highest frequency in West Africa.'

Marla expressed that the results were "deeper" than she had wanted and referred to ancestry "far before the time that I am interested in." She continued, expressing frustration that these genetic genealogy test results did not provide her with more information than she might have surmised on her own:

> Huh???? Ethiopia? West Africa? Didn't just about everybody outside Africa come through the Ethiopia area 50,000 years ago? Maybe I'm off by a few thousand years . . . . These kinds of results are meaningful for those tracking the worldwide movement of people (like the National Geographic study), but not really meaningful to me in my much narrower focus.

Marla concluded our exchange by informing me of her plan to send these results to the African Ancestry company for re-interpretation and comparison against its ethnic lineage database. Washington, D.C.-based African Ancestry was founded by geneticist Rick Kittles and his business partner Gina Paige in 2003. The company sells two genetic genealogy tests, MatriClan and PatriClan, which analyze mtDNA- and Y-DNA-linked genetic information, respectively, in order to associate customers with present-day African ethnic groups based on matches with its proprietary biobank, the African Lineage Database (ADL). As I described, African Ancestry's services are popular among root-seekers of African descent because they approximate Haley's *Roots* journey and because the company claims to hold the largest collection of "African DNA."

Cecily and Marla's comments indicate that effectual test outcomes are those that offer test-takers a *usable* past. For Cecily, composite testing would merely confirm the ancestral hybridity (resulting from racial slavery) of which she was already convinced; to her mind, this form of scientific genealogical analysis provided information that was neither novel nor useful. Given Marla's aim to derive ethnic lineage from spatio-temporal results, the "much narrower focus" that would be "really meaningful" to her would apparently take the form of a genetic genealogy that affiliated her with an African ethnic group and possibly a present-day nation, thus fulfilling the genealogical aspiration that was established when she attended the Alex Haley presentation three decades earlier. Taken together, Cecily's indifference toward racio-ethnic composite testing and Marla's preference for ethnic lineage testing suggest that not just any scientific evidence of ancestry will do. Rather, consumers come to genetic genealogy testing with particular questions to be answered, with mysteries to solve, with personal and familial narratives to complete and seek "the right tools for the job."[31] Genetic genealogy test results may challenge not only prior expectations but also other evidentiary bases of self-perception and social coherence. As Marla's response to the spatio-temporal result implies, and as I elaborate below, genetic genealogists are judicious not only about the types of genetic genealogy tests they purchase, but also about the significance of the test results they receive. The negotiation of test results throws into relief the inadequacy of the race naturalist – race pragmatist contest. For, through their efforts to align *bios* with bios, root-seekers selectively imbricate discrete epistemologies of race and ethnicity.

Genealogical Disorientation: "We Still Technically Don't Know Who we Are"

I attended a symposium on race and genetics at a large public urban university in the Midwest in the fall of 2003. It was a small, interdisciplinary gathering of schol-

[31] See Adele E. Clarke and Joan Fujimura.

ars and included presentations by social scientists, geneticists, and bioethicists, among others. The audience consisted mostly of symposium presenters, but also included interested faculty affiliated with the university and members of the public, who sat-in on discussions for short periods of time throughout the day. A smattering of non-academics was on-hand for an afternoon presentation by Rick Kittles, African Ancestry's chief science officer and, at the time, also a director at the National Human Genome Center at Howard University. In a manner that blended erudition and affability, Kittles discussed the scientific research and socio-cultural assumptions behind the ethnic lineage analysis his company had begun offering several months prior.[32] During the presentation, I sat next to a middle-aged African American woman, whose cotton navy jacket emblazoned with Teamsters Union patches and steel-toed work boots placed her in a somewhat different genealogy of interest in the topic than the academics in attendance, who, like me, were dressed in business casual attire and hunched over our notebooks. While Kittles continued his dynamic performance, the woman nodded enthusiastically in assent and, from time to time, looked over to me seeking mutual appreciation of the geneticist's presentation. I smiled and nodded in return. This silent call-and-response went on for several minutes when at one point she leaned in and whispered to me that she had "taken his test."

At the conclusion of Kittles' presentation, the woman (Pat) and I continued our discussion of her experience with African Ancestry's genetic genealogy service as she strode with me through the labyrinthine campus – less confusing to her as a university employee – to the public transport stop from which I would travel to the airport. As we walked and talked, she spoke of her interest in conventional genealogy and of recent events that had prompted her to use DNA analysis to trace her African roots. Pat told me that she is a longstanding member of the AAHGS and of two other genealogical societies. For almost 30 years she had assembled archival materials, reminiscences, oral history, and linguistic clues from family members. This evidence led her to deduce that her family's maternal line may have descended, in her words, from "the Hottentots" (or the Khoisan of southern Africa). Despite some success with her genealogical research by traditional means, Pat has not been able to locate a slave-ship manifest or definitive documentation of her African ancestry. She told me that, as a result, "some missing links" remained to be uncovered for her.[33]

[32] In addition to the fact that African Ancestry supplies genetic genealogists with the particular tools required to re-trace their own versions of Alex Haley's roots journey, its success and popularity among blacks owes as well to Kittles' "authentic expertise" as an African American and a geneticist. For more on this, see Alondra Nelson.

[33] Pat's phrase "missing link" suggests that she feels that integral knowledge about her ancestral background remains unknown, but also calls to mind discourses of human evolution.

Prior to Pat's employment at the university, she processed forensic evidence for a police department crime lab in the same city. This work experience bolstered her confidence in African Ancestry's product. As she explained to me, "I've seen people let off jail sentences based on DNA . . . I'm not question[ing] about DNA . . . given my experiences [working in the lab], there is no reason to doubt the technology." Pat's resolute belief in genetic analysis was paralleled by her faith in the company's black chief scientist. At our initial meeting, she unequivocally said, "I trust Dr. Kittles." Owing to a legacy of racially segregated healthcare and experimental exploitation, such as the notorious Tuskegee syphilis study, some African Americans are mistrustful of bioscientific research. This mistrust has been shown to negatively impact contemporary health-seeking behavior and to create a disincentive for African Americans to participate in clinical trials.[34] In contrast, the growing popularity of African Ancestry's services demonstrates that this long-held skepticism may be overcome by the trustworthy persona of a researcher-entrepreneur combined with a passion for root-seeking, and in Pat's case, in particular, familiarity with genetic analysis and its usefulness in exonerating wrongly incarcerated persons.[35]

Pat purchased an mt-DNA test from African Ancestry following a persuasive pitch by Kittles at one of the three genealogy club meetings she regularly attends. Asked to recall her feelings as she awaited her genealogy test results, she responded, "I didn't know what to expect…it's like rolling a lottery thing, okay this is where it landed." By this, Pat meant that she came to the testing with prior information but not necessarily pre-conceptions about her ethnic "match." A comparison of Pat's DNA with the ADL did not place her maternal line in southern Africa. Instead, she was associated with the Akan, a large ethnic group of Ghana and south-eastern Cote D'Ivoire that includes the Asante, the Fante, and the Twi, among others. Pat's results included "my genotype" printed out on paper, "a letter of authenticity from the lab," and "a certificate saying I was Akan." However, these authoritative artifacts did not leave Pat feeling settled about her ancestry. She recollected, "I felt numb, blank. [I've] been doing genealogy since 1977. I grew up with knowledge of Hottentot . . . all these years later, I find out its Ghana." She added, after a pause, "What if it's true?"

Pat's uncertainty about her results increased several weeks later, when she learned that other members of her genealogy club reported receiving the same eth-

[34] See Giselle Corbie-Smith; Vanessa Northington Gamble.

[35] Indeed, many of the most highly publicized cases of wrongly accused persons being exonerated by DNA analysis have involved African Americans. The emancipation of blacks in these instances may mitigate the negative legacy of the Tuskegee study. In the UK, many of the black Britons who participated in the *Motherland* study were apprehensive about providing their DNA because of concerns about privacy (interview with Arthur Torrington November 20, 2005).

nic match from African Ancestry as she did – Akan. These results may be accurate in the statistical universe of gene sequence variants, especially in light of substantial historical research showing that current-day Ghana and other western African countries were key nodes in the trans-Atlantic slave trade. Nevertheless, the preponderance of similar ethnic lineage findings among her genealogist colleagues, and the inconsistency of her genetic result with the family genealogy she had laboriously assembled by conventional means, led Pat to conclude that "we still technically don't know who we are."

Pat's use of the word "technically" in her estimation that "we still . . . don't know who we are" is of signal interest. It indicates that although she had expected that genetic genealogy testing would provide her with a family of origin in Africa, the results did not fully convince her. She harbors some doubt about the reliability of the techniques used. However, given her prior positive assessment of genetic testing, her reference to technical uncertainty also seems to indicate discomfort with conceptualizing family history as a technical matter. On one level, her uncertainty might be interpreted as an instantiation of the "genealogical dis-ease" that may result from the collision of kinship concepts and genetic science.[36] Yet, her words intimated that the genetic genealogy testing experience produced something else as well, a lack of orientation, and more particularly, "genealogical disorientation" as an affect ("I felt numb") and as an effect of her misgivings about its reliability ("What if it's true?"). Pat also feels "blank." In her search for family, she has lost the familiar. It was not evident during this conversation if the reinscription that these feelings of "blankness" might initiate would produce a deepened investment in how she perceived of herself prior to the testing experience, would catalyze a new inscription of genetic deterministic thinking about her racio-ethnic identity, or produce some combination of these.[37] For Pat and others with whom I have spoken, the receipt of genetic facts about ancestry opened up new questions about identity and belonging, rather than settling them absolutely.[38] The test results did impart a topographical orientation to Pat in the form of an abstract matrilineal link to the Akan that consequently associated her with a country in Africa. But the DNA analysis failed to orient her in a more phenomenological sense. It did not orient her *towards* the social world in a particular manner, nor did it give her bearings *in relation to* other persons or collectivities.

[36] See Rayna Rapp, Deborah Heath and Karen-Sue Taussig.

[37] In my analysis of Pat's experience, reinscription is distinct from the "molecular reinscription of race," Duster's description of the use of "patterns in DNA for 'predicting' ethnic and racial membership" in pharmacogenetics and forensic science (427). However, reinscription is not unrelated re-writings. Pat's "blankness" was prompted by the molecular or genetic logic to which Duster refers.

[38] See Carl Elliot and Paul Brodwin.

Since receiving her test results, Pat has endeavored to fashion this second, phenomenological sense of orientation for herself. She has begun a friendship with a Ghanaian neighbor and has embarked on research into the history and culture of the Akan. More recently, Pat has begun to explore the possibility of having roots in West Africa. Yet her DNA test has taken on deeper significance, not because of increased confidence in mt-DNA analysis, but because of her own efforts to resolve her genealogical disorientation. The growing appeal of Pat's Akan-ness was powerfully illustrated in the following account she shared with me: at a community Kwanzaa fair, Pat was faced with a purchasing decision that revealed her vacillating racio-ethnic identity. Coming upon an African immigrant flag-vendor as she strolled through the fair, Pat was confronted with two symbols of her ancestral roots and putative nationality. She inquired about the significance of a flag with three fields of red, black, and green. The vendor replied that it was a "general flag," indicating that it was a pan-African flag, which symbolized the African diaspora rather than a specific nationality or ethnicity. Pat responded, "My DNA said I came back as Ghanaian. I don't need the red, black, and green." The woman replied, "Now you know, so you don't need just a plain flag anymore." She concluded that "if anything has changed [about how I perceive myself], it's that I bought my first Ghanaian flag last year." In this exchange, Pat's testing experience emboldened her to invoke her *soma* ("my DNA said I came back as Ghanaian") when offered an undifferentiated symbol of Africa by the vendor. She then asserts that she may not "need" the pan-African flag. However, it is the African vendor's not uninterested response, "Now you know," that endorses and authenticates Pat's claim to Ghana, leading to the purchase of a symbol of her possible "home."[39] Although this social interaction encounter authorized Pat's genetic affiliation with the Akan, her opinion of her family origins nevertheless remains in flux. Now when asked by others about the outcome of her root-seeking pursuits, Pat admits to answering "Akan" and "Hottentot" interchangeably.

[39] To be sure, Pat and other test-takers exercise consumer choice in the interpretation of their genetic test results. On the one hand, the particular freedom to choose, links her purchase of ethnic lineage testing to her purchase of the Ghanaian flag. On the other, her exchange with the vendor suggests that her choice is somewhat constrained by the African woman's cultural authority, just as it is by African Ancestry's authenticating but limited biobank. For discussions of the consumption of technology, see Nelly Oudshoorn and Trevor Pinch; and Alondra Nelson and Thuy Linh N. Tu.

Reckoning African Ancestry: Affiliative Self-Fashioning and Genetic Kinship

In *Picturing Personhood,* anthropologist Joseph Dumit explores the interplay between technologies that make the inner-workings of the brain visible and "biomedical identity." Brain images and genetic genealogy tests are very different inscriptions of scientific knowledge, but they are similarly metonymic. Both are artifacts of bioscience that circulate beyond the laboratory and the clinic. Brain scan images and certificates announcing genetically derived race or ethnicity are "received-facts" that can be incorporated into a process of subject formation that Dumit calls "objective self-fashioning." He defines this as "an ongoing process of social accounting to oneself and others in particular situations in which received-facts function as particularly powerful resources because they bear the objective authority of science" (44). Applied to the geneticization of race and ethnicity, the concept of objective self-fashioning elucidates how individual and collective identities are constituted through both extant frameworks (for example, historical knowledge, collective memory, conventional genealogy, and alternative genetic kinship accounts) and novel techniques.[40] But as Pat's experience illustrates, and as I detail below, with genetic genealogy testing, test-takers enact a course of deliberate and strategic negotiation in an effort to create kinship orientation that is not taken account of in Dumit's analysis. Therefore, extending Dumit's analysis, I term this process "affiliative self-fashioning."[41] Whereas objective self-fashioning highlights the epistemological authority of received-facts that become resources for self-making at specific "location[s] of social accounting" (Dumit 44) such as the lab or the courtroom, affiliative self-fashioning attends as well to the weight of individual desires for relatedness, for "communities of obligation" (Rose 177) and how this priority shapes evaluations of the reliability and usability of scientific data. In the process, received-facts are also reconciled with a complex of alternative identificatory resources.

I attended a symposium on race and genetics at the London School of Economics in 2004. Neil Cameron, one of the producers of the 2003 BBC documentary, *Motherland: A Genetic Journey* also took part in this two-day meeting. At the conclusion of the first day's discussions, he invited me to a gathering of participants in the study on which the documentary was based and I eagerly accepted his invitation. I arrived at the venue, the Museum of London, to find about a dozen men and

[40] Also see Keith Wailoo and Stephen Pemberton.

[41] 'Self-fashioning' was first elaborated by Stephen Greenblatt with reference to Renaissance aesthetics in Renaissance Self-Fashioning: From More to Shakespeare. I am not the first to suggest other genres of self-fashioning: in addition to Joseph Dumit's "Is it Me or My Brain" and Picturing Personhood, see also Nadia Abu El-Haj on 'territorial self-fashioning' in the Middle East in Facts.

women assembled in a small seminar room. This was a gathering of the Motherland Group, black Britons who participated in a study commissioned for the documentary for which they had volunteered DNA samples in exchange for the opportunity to have their ancestral links to Africa scientifically established. The Motherland Group first convened in March 2003, shortly after the premiere of the documentary and at the request of a few study participants. According to Arthur Torrington, a participant in the study and leader of the Motherland Group, members came together to stimulate discussions that would make the "pros and cons of the testing clear" because, as he expressed it, the test results were but "the beginning of a journey; there is much more to this thing." Producer Neil Cameron, who manages the study's data protection registry, contacted participants. He and his production partner, Archie Baron, subsequently arranged presentations to the group by genetics experts. As Cameron explained to me, group members "wanted to be able to talk to each other about the experience and learn more about the science behind the study."

This meeting served as a forum for the co-production of biological and social identities, for the making of what Paul Rabinow has called "biosociality." However, participants did not gather on the basis of what have now become canonical examples of biosociality – shared medical conditions, genetic predisposition to disease or disability.[42] Rather, as both Cameron's and Torrington's comments about the purpose of the Motherland group suggest, rather than on the basis of accepted or determined biological identities, members assembled to explore what biosociality might result from their testing experience. This was *bios* put to the task of creating a very particular kind of sociality: the possibility of kinship based on ethnic lineage and ethno-racial composite DNA analysis. It was a venue for the constitution of individual and collective identities through both objective and affiliative self-fashioning.

Although DNA samples from 229 persons were analyzed for the Motherland study, the documentary featured just three participants, chosen by the producers for their telegenic appeal and for the dramatic potential of their narratives. Cameras accompanied Jacqueline Harriot as she traveled to Jamaica to explore her more recent Caribbean heritage and await the results of her racio-ethnic composite analysis.[43] *Motherland* viewers traveled as well with Mark Anderson and Beaula McCalla as they were transported to their supposed, respective pre-slave trade

[42] See Rayna Rapp, Deborah Heath and Karen-Sue Taussig on achondroplasia; Nikolas Rose and Carlos Novas' work for further reading on biological citizenship; and also Sahra Gibbons and Carlos Novas.

[43] Harriot's test determined her racio-ethnic composite result to be 28% European and 72% sub-Saharan African.

"motherlands" of Niger and Bioko Island (an island of Equatorial Guinea) for a dramatic "reunion" with their lost kin.

Beaula was in attendance at this meeting of the Motherland Group, which featured human evolutionary geneticist, Martin Richards, a specialist from the University of Leeds in African migration, as the day's speaker. His presentation outlined the theoretical and technical assumptions on which the study participants' genetic genealogy results rested, and cautioned attendees about the limitations of mtDNA and Y-chromosome analyses for the purposes of determining ancestry. Specifically, Richards warned attendees that genetic genealogy does not link a consumer to ancestors at a specific place and time. He stressed as well that the proprietary DNA databases on which genetic genealogical tests rely are incomplete, because they contain too few samples from too few sites in Africa to make robust claims about any given individual's ancestry.[44]

On these points, Richards was repeating arguments first aired in an editorial critique of the Motherland study that he published in the *Guardian* newspaper of London 1 year earlier. Entitled "Beware the Gene Genies," his opinion piece revealed that the genetic marker used by Cambridge University geneticist Peter Forester to link Beaula with the current day Bubi people of Bioko Island, Equatorial Guinea—called a "rare marker" in the documentary—was also found thousands of miles away. Richards wrote that "a glance at the published mitochondrial database shows that Beaula's variant is also found in Mozambique." He continued, noting that "a huge area of central and southern Africa that provided more than a third of all victims of the slave trade is still unsampled. Beaula's maternal lineage could have come from anywhere in that region." I was later told by producer Cameron that Motherland Group members were familiar with Richards' editorial. Nonetheless, Richards' repeated criticisms of the project on this day supplied a moment of quiet tension in the room – expressed through hushed "tsk-tsks" and sheepish glances. I initially attributed this tension to the fact that Beaula (who I recognized from the documentary) was present as the geneticist delineated the two possible, but possibly conflicting, accounts of her ancestry. However, at the conclusion of Richards' talk, it became clear that Beaula was not the only person in the room who might be experiencing genealogical disorientation.

During the question and answer period, a self-described "Grenadian-born British" woman, who appeared to be in her late fifties and who I will call Delores, announced that in addition to herself, "twelve or fifteen women came up Bubi in the [Motherland] study" and concluded that the test "does not give a complete blueprint of who I am." Delores' evaluative rhetorical posture established her role as the interpreter of her genetic genealogy test results and underscored the process of affiliative self-fashioning in which she was engaged at this gathering. Her use of

[44] Also see Charles Rotimi; and Robert Ely et al.

the word "blueprint" may reflect some degree of accord with a geneticized conception of her ancestry.[45] Like Pat's use of the word "technically," however, the qualifier "complete" here seemed to express a conviction that the genetic genealogy test offered only a partial account of her identity. Delores' knowledge of the "twelve or fifteen" other "Bubis" in the study also undermined her sense of individuality – her desire for "a complete blueprint of who *I* am," as she put it. The genetic findings did not seem to satisfy her criteria for either genetic or social exclusivity.

Many who attended the Motherland gathering stayed after Richards' presentation to ask additional questions and socialize over tea and cookies. After hearing Delores voice her concerns, I was curious about Beaula's opinion of the second possible interpretation of her genetic genealogy test. Did she, like Delores, question the test's reliability because of the presence of the "rare" Bubi marker in several other Motherland study participants? Which result, if any, did she accept after learning that her ancestry might also be traced to south-eastern Africa? On what basis did she decide between the alternative accounts of her maternal lineage? I introduced myself to Beaula and we began a conversation about her experience as a participant in the *Motherland* documentary project. Before she could fully respond to my queries, a man who had been sitting next to her at the meeting joined us. She introduced him to me: "This is my brother, Juan. He doesn't speak English. He speaks Spanish [an official language of Equatorial Guinea]." "Your biological brother?," I asked. "My brother from Equatorial Guinea," she responded. From this point on, the discussion continued between the three of us, with me alternating between elementary Spanish with Juan, and English with Beaula (who spoke even less Spanish than me), but drifted from the topic of Beaula's genetic ancestry tracing, to the purpose of my visit to London and Juan's impressions of England. As a consequence, I was not able to inquire further about Beaula's brief, but suggestive, statement of affiliation with Juan on that day. Nevertheless, because I was subsequently in contact with Beaula and Juan by email, I can make a few provisional observations about this exchange and what it suggests about which ancestry association Beaula finds most compelling.

When posing the question, "Your biological brother?," I wanted to know how she defined her relationship with Juan. Although she is depicted in *Motherland* as living an Afro-centric lifestyle in Bristol, England, her classification of Juan suggested to me a relationship that was more significant than the vernacular term "brother" used by some blacks to refer to others of African descent. Moreover, in an email, Juan informed me that "Beaula esta ayudando a encontrar a mi madre [Beaula is helping me find my mother]" – a mother with whom he had lost contact

[45] Her use of the word "blueprint" may reflect some degree of accord with a geneticized conception of her ancestry. Abby Lippman elaborates on this notion of geneticization in "Prenatal Genetic . . . ."

many years ago, and who may have migrated from Equatorial Guinea to Europe. This statement suggested that Beaula and Juan did not share a mother, nor was it likely that they were members of the same nuclear family.[46] If Beaula did not regard Juan as a relation in either of these two senses, perhaps "my brother from Equatorial Guinea" described what Catherine Nash terms "genetic kinship," affiliations fashioned from the facts of DNA analysis, the particulars of which are both unspecified and ahistorical.[47] What was certain was that there was something that linked Beaula to Juan and, moreover, obligated her to assist him with his own familial search.[48] Sponsorship of an orphanage and school on Bioko Island and the cultivation of a growing number of Equatoguinean acquaintances both in Africa and Europe also indicated Beaula's chosen affinity. Her assertion of a familial and ethnic tie to Juan confirmed that, whatever her feelings about the genetic link to Mozambique, she was committed to the ancestry designation she received as a participant in the Motherland study. Similar to Pat's experience, Beaula's genetic genealogy result amassed traction through social ties.[49]

Racial and Ethnic Projects: Site, Scale, and Subjectification

Almost weekly, it seems that media outlets in the USA and the UK publish dramatic accounts of genealogical disorientation experienced by persons whose genetic analyses reveal new or surprising information about their ancestry. Among this genre of unexpected and unsettling kinship narratives are stories of persons whose presumed racial and ethnic identities are overturned by genetic testing, including the University of Miami professor who was said to be related to Genghis Kahn based on Y-DNA analysis, several persons in Yorkshire, England, who were found to share a Y-chromosome haplotype uncommon in Europe that ancestrally linked them to Africa, and a black-identified civil rights advocate in California whose DNA was said to exhibit no trace of sub-Saharan African ancestry.[50] On the

[46] It is worth noting here that Juan has "lost his mother," to paraphrase Saidiya Hartman's poignant exploration of African American root-seeking and its psychoanalytic implications. In this regard, he was a root-seeker like Beaula and the others discussed in this paper.

[47] Genealogists also use the phrase "DNA cousin" to characterize persons who might share a set of genetic genealogy test markers but whose relationship to one another remains unspecified. In this instance, "DNA" marks filial ambiguity *and* scientific precision, see Nelson's "The Factness of Diaspora."

[48] As I discuss in another paper ("The Factness of Diaspora") the generation of responsibilities and rights and forms of exchange across the Africa diaspora is a common end-result of genetic genealogy testing. In this way, genetic genealogy testing facilitates the formation a spatial, interactional diasporic network. See also Laurence J. C. Ma.

[49] See David M. Schneider.

[50] See Nicholas Wade; "Yorkshire Clan Linked to Africa;" and Erin Aubry Kaplan.

other end of the affective spectrum are published reports of African Americans who, following the use of genetic genealogy testing, claim finally to know who they are. These roots narratives follow a now predictable arc: DNA testing, feelings of completion, and the assumption of the subject's unwavering confidence in the genetic test outcome. Press accounts such as these leave little doubt that genetic truth of identity and kinship will out, that social categories such as race and ethnicity are being made anew from the whole cloth of As, Cs, Gs and Ts.

Away from the glare of the media, however, test-takers can exercise latitude in determining the import of genetic ancestry analysis. To be sure, root-seekers come to genetic genealogy with the expectation that it will supply definitive information about their family histories. But a genetic "match" is just the beginning of a process of identification, rather than its conclusion. After test results are rendered, root-seekers endeavor to translate them from the biological to the biographical, from a pedigree of origins to a satisfying life story. They attempt to meaningfully align *bios* with bios.

The efforts to reconcile "nature" and "culture" into identity that commence following the use of genetic genealogy testing call into question race naturalist and race pragmatist polemics. More analytic purchase can be gained by conceiving of contests over the "reality" of race as contrasting, but ultimately interlinked, "racial projects": social ventures through which race and ethnicity are fashioned, interpreted, represented, and institutionalized.[51] From this perspective, naturalism and pragmatism come into view as but two of many historical and coeval projects. Attentiveness to this multi-layered process should re-direct scholarly focus away from zero-sum definitional boundaries and toward issues of *scale*, *site*, and *subjectification*.

Race is constituted in the law, the media, bioscience, and from seemingly small acts that occur at "the level of everyday life" as well (Omi and Winant 56). Racial projects are thus scalar; they take form at the interplay of macro- meso- and micro-level processes. As the experiences of genetic genealogy test-takers reveal, the constitution of race and ethnicity may fluctuate by scale – the codification of social categories in commercial laboratories, for example, is not always or easily translated into the geneticization of identity. Accordingly, analyses should traverse levels of scale from the microscopic, byte-sized "molecularizaion" of race, through the individual and collective lived experience of social identity, and to large-scale racialization and ethnicization.[52]

It is also important for scholars to take notice of the *specificity* of scale, or site. In drawing attention to site, I seek to highlight the significance of where—in what

[51] See Omi and Winant.

[52] See Nikolas Rose; Duana Fullwiley, "The Molecularization;" and Steven Epstein (both works).

social and physical situations—race and ethnicity may be constituted for the robustness of scholarly claims. Although some generalizations can be made about the pervasiveness of geneticization, there are distinctions that are also worth noting. In the domain of genetic testing, similar forms of DNA analysis produce evidence at specific locations that differently permit or constrain identification. For example, mtDNA analysis is used not only in genealogy testing, but also in medical genetics (for example, research into type 2 diabetes)[53] and forensic science (when there is a small sample of biological evidence). However, as I have described, commercial mtDNA analysis for genealogical purposes supplies racio-ethnic lineage information that may permit consumer interpretation, while mtDNA testing in a criminal justice setting, in which the genetic "matching" criteria is narrowly delimited and backed by state authority, leaves little space for negotiation. Taking account of site variance such as this can yield richer insights about both racialization and geneticization.

Subjectification—subjects making themselves and being made by forces beyond themselves—is related to site and is also important to discussions of race and genetics. Genetic genealogy testing opens up "ethnic options" to blacks in the USA and the UK that may have been previously unavailable.[54] However, the affiliative self-fashioning it may spur is enacted from within what might be understood the "iron cage" of the genome.[55] The testing promises to reveal elusive knowledge, yet the particular longings that root-seekers of African descent seem to feel when they resort to it are shaped by distinct histories of slavery and the continuing realities of racial oppression. Root-seekers' sense of autonomy and empowerment may come at the cost of acquiescence to a classificatory logic of human types that compounds, rather than challenges, social inequality. The affiliative self-fashioning described here is then a limited type of agency, unfolding from within less mutable social structures.

Examining genetics and race from the perspectives of scale, site, and subjectification alerts us to the intricacies of the "genome-in-practice."[56] The decoding of the human genome does not and cannot verify the validity of race because many epistemologies and ontologies contribute to its meaning and significance. It is too early to assess what long-term effects genetic genealogy testing may have on social norms or political affiliations. Yet, it is apparent that the practice of crafting racial and ethnic subjects through genetic genealogical designation—what has been cava-

[53] Research by S. W. Ballinger et al., for instance, demonstrates the use of mtDNA in such studies.

[54] See Mary Waters' texts.

[55] See Joan H. Fujimura. However, as Fujimura rightly suggests, the "iron cage" is not intractable. Root-seekers' perspectives give them potential, and possibly powerful, agency in the process of translating, contesting and transforming science.

[56] See Alan Goodman, Deborah Heath and M. Susan Lindee.

lierly described as "recreational" genomics—displays complexity that warrants notice and further investigation.

Note

The research for this paper was supported by a Career Enhancement Fellowship from the Woodrow Wilson and Andrew S. Mellon Foundations; a Postdoctoral Diversity Fellowship from the Ford Foundation; and the Social Science Research Fund, Yale University. I thank Nicole Ivy for editorial assistance,

Works Cited

American Anthropological Association. "American Anthropological Association Statement on "Race." 1998. 17 September 2003 <www.aaanet.org/stmts/racepp.htm>.

American Sociological Association. "The Statement of the American Sociological Association on the Importance of Collecting Data and Doing Social Scientific Research on Race." 2002. 10 October 2003 <www.asanet.org/medi.asa_race_statement.pdf>.

Baker, Lee. *From Savage to Negro: Anthropology and the Construction of Race, 1896-1954*. Berkeley, CA: University of California Press, 1998.

Ballinger, Scott W. et al. "Maternally Transmitted Diabetes and Deafness Associated with a 10.4 kb Mitochondrial DNA Deletion." *Nature Genetics* 1.1 (1992): 11-15.

Bamshad, Michael J. et al. "Human Population Genetic Structure and Inference of Group Membership." *American Journal of Human Genetics* 72.3 (2003): 578-589.

Burchard E. Gonzalez et al. "The Importance of Race and Ethnic Background in Biomedical Research and Clinical Practice." *New England Journal of Medicine* 348.12 (2003): 1170-1175.

Cann, Rebecca, Mark Stoneking, and Allan Wilson. "Mitochondrial DNA and Human Evolution." *Nature* 325 (1987): 31-36.

Cavalli-Sforza, Luigi, Paolo Menozzi, and Alberto Piazza. *The History and Geography of Human Genes*. Princeton, NJ: Princeton University Press, 1994.

Clarke, Adele E. and Joan Fujimura (eds.). *The Right Tools for the Job: At Work in Twentieth-Century Life Sciences*. Princeton, NJ: Princeton University Press, 1992.

Condit, Celeste Michelle. *The Meanings of the Gene: Public Debates about Human Heredity*. Madison, WI: University of Wisconsin Press, 1999.

Cooper Richard S., Jay S. Kaufman, and Ryk Ward. "Race and Genomics." *New England Journal of Medicine* 348.12 (2003): 1166-1170.

Corbie-Smith, Giselle. "The Continuing Controversy of the Tuskegee Syphilis Study: Implications for Clinical Research." *American Journal of Medical Sciences* 317.1 (1999): 5-8.

Dumit, Joseph. *Picturing Personhood: Brain Scans and Biomedical Identity.* Princeton, NJ: Princeton University Press, 2003.

---. "Is It Me or My Brain: Depression and Neuroscientific Facts." *Journal of Medical Humanities* 24.1.2 (2003): 35-47.

Duster, Troy. "The Molecular Reinscription of Race: Unanticipated Issues in Biotechnology and Forensic Science." *Patterns of Prejudice* 40.4.5 (2006): 427-41.

---. *Backdoor to Eugenics.* 2nd ed. New York: Routledge, 2003.

El-Haj, Nadia Abu. "'A Tool to Recover Past Histories': Genealogy and Identity After the Genome." *Occasional Papers of the School of Social Science.* Institute for Advanced Study, Princeton, New Jersey, 2004 <www.sss.ias.edu/publications/papers/paper19.pdf>

---. *Facts on the Ground: Archeological Practice and Territorial Self-Fashioning in Israeli Society.* Chicago, IL: University of Chicago Press, 2002.

Elliot, Carl and Paul Brodwin. "Identity and Genetic Ancestry Tracing." *British Medical Journal* 325 (2002): 1469-1471.

Ely, Robert et al. "African-American Mitochondrial DNAs Often Match mtDNAs Found in Multiple African Ethnic Groups." *BMC Biology* 4.34 (2006): 1-25.

Epstein, Steven. *Inclusion: The Politics of Difference in Medical Research.* Chicago, IL: University of Chicago Press, 2007.

---. "Bodily Differences and Collective Identities: The Politics of Gender and Race in Biomedical Research in the United States." *Body and Society* 10.2.3 (2004): 183-203.

Fausto-Sterling, Anne. "Refashioning Race: DNA and the Politics of Health Care." *differences: A Journal of Feminist Cultural Studies* 15.3 (2004): 1-37.

Fullwiley, Duana. "The Biologistical Construction of Race: 'Admixture' Technology and the New Genetic Medicine." *Social Studies of Science* 00 (2008): 000-000.

---. "The Molecularization of Race: Institutionalizing Racial Difference in Pharmacogenetics Practice." *Science as Culture* 16.1 (2007): 1-30.

Fujimura, Joan H. "Sex Genes: A Critical Sociomaterial Approach to the Politics and Molecular Genetics of Sex Determination." *Signs: Journal of Women in Culture and Society* 32.1 (2006): 49-82.

Gamble, Vanessa Northington. "Under the Shadow of Tuskegee: African Americans and Health Care." *Tuskegee's Truths: Rethinking the Tuskegee Syphilis Study.* Ed. Susan Reverby. Chapel Hill, NC: University of North Carolina Press, 2000. 431-442.

Gibbons, Sahra and Carlos Novas. *Biosocialities, Genetics and the Social Sciences: Making Biologies and Identities.* London: Routledge, 2007.

Gilroy, Paul. *Against Race: Imagining Political Culture Beyond the Color Line.* Cambridge, MA: Harvard University Press, 2000.

Goldberg, David Theo. *Anatomy of Racism.* Minneapolis, MN: University of Minnesota Press, 1990.

Goodman, Alan, Deborah Heath, and M. Susan Lindee (eds.). *Genetic Nature/Culture: Anthropology and Science Beyond the Two-Culture Divide.* Berkeley, CA: University of California Press, 2003.

Greenblatt, Stephen. *Renaissance Self-Fashioning: From More to Shakespeare.* Chicago, IL: Chicago University Press, 1980.

Hacking, Ian. "Why Race Still Matters." *Daedalus* 134.1 (2005): 102-116.

Haley, Alex. *Roots: The Saga of an American Family.* New York: Dell, 1976.

Hammer, Michael F. "A Recent Common Ancestry for Human Y Chromosomes." *Nature* 378.6555 (1995): 376-378.

Haraway, Donna. *Modest_Witness@Second_Millenium.FemaleMan©_Meets_Oncomouse&tm: Feminism and Technoscience.* New York: Routledge, 1997.

---. *Simians, Cyborgs, and Women: The Reinvention of Nature.* New York: Routledge, 1991.

Hartman, Saidiya. *Lose Your Mother: A Journey Along the Atlantic Slave Route.* New York: Farrar, Straus and Giroux, 2007.

Heath, Deborah et al. "Nodes and Queries: Linking Locations in Networked Fields of Inquiry." *American Behavioral Scientist* 43 (1999): 450-463.

Helmreich, Stefan. "Spatializing Technoscience." *Reviews in Anthropology* 32.1(2003):13-36.

Hine, Christine M. *Virtual Ethnography.* London: Sage Publications, 2002.

Jacobson, Matt. *Whiteness of a Different Color: European Immigrants and the Alchemy of Race.* Cambridge, MA: Harvard University Press, 1982.

Jobling, Mark A. and Chris Tyler-Smith "Fathers and Sons: The Y Chromosome and Human Evolution." *Trends in Genetics* 11.11 (1995): 449-455.

Kaplan, Erin Aubry. "Black Like I Thought I Was." *LA Weekly* 7 October 2003 < http://www.laweekly.com/2003-10-09/calendar/black-like-i-thought-i-was>.

Lewontin, Richard C. *Human Diversity.* San Francisco, CA: Scientific American Books, 1982.

Lippman, Abby. "The Politics of Health: Geneticization Versus Health Promotion." *The Politics of Women's Health: Exploring Agency and Autonomy.* Ed. Susan Sherwin. Philadelphia, PA: Temple University Press, 1998. 64-82.

---. "Prenatal Genetic Testing and Screening: Constructing Needs and Reinforcing Inequalities." *American Journal of Law and Medicine* 17.1.2 (1991): 15-50.

Lock, Margaret. "The Eclipse of the Gene and the Return of Divination." *Current Anthropology* 46.5 (2005): 47-70.

Lopez, Ian Haney. *White By Law: The Legal Construction of Race*. New York: New York University Press, 1996.

Ma, Laurence J. C. "Space, Place, and Transnationalism in the Chinese Diaspora." *The Chinese Diaspora: Space, Place, Mobility, and Identity*. Ed. Laurence J. C. Ma and Carolyn Cartier. Lanham, MD: Rowman and Littlefield, 2003. 1-50.

McCalla, Beaula. "Finding Me, an Account of a Genetic Journey to Afrika." Port-Cities Bristol. 21 March 2005 <www.discoveringbristol.org.uk/showNarrative.php?sit_id=1&narId=557&nacId=558.

Miller, Daniel and Don Slater. *The Internet: An Ethnographic Approach.* Oxford: Berg Publishers, 2001.

Moore, Donald S., Anand Pandian and Jake Kosek. *Race, Nature, and the Politics of Difference*. Durham, NC: Duke University Press, 2003.

Motherland: A Genetic Journey. Dir. Archie Baron. British Broadcasting Corporation (BBC2). 14 February 2003.

Nash, Catherine. "Genetic Kinship." *Cultural Studies* 18.1 (2004): 1-33.

Nelson, Alondra. "The Factness of Diaspora: The Social Sources of Genetic Genealogy." *Revisiting Race in a Genomic Age*. Ed. Barbara Koenig, Sandra Soo-Jin Lee, and Sarah Richardson. New Brunswick, NJ: Rutgers University Press, 2008. 453-483.

Nelson, Alondra and Thuy Linh N. Tu (eds.). *Technicolor: Race, Technology and Everyday Life.* New York: New York University Press, 2001.

Omi, Michael and Howard Winant. *Racial Formation in the Unites States: From the 1960s to the 1990s*. 2nd ed. New York: Routledge, 1994.

Ossorio, Pilar and Troy Duster. "Race and Genetics: Controversies in Biomedical, Behavioral, and Forensic Science." *American Psychologist* 60.1 (2005): 115-128.

Oudshoorn, Nelly and Trevor Pinch. "Introduction: How Users and Non-Users Matter." *How Users Matter: The Co-Construction of Users and Technology*. Ed. Nelly Oudshoorn and Trevor Pinch. Cambridge, MA: MIT Press, 2005. 1-27.

Rabinow, Paul. *Essays on the Anthropology of Reason.* Princeton, NJ: Princeton University Press, 1996.

Rapp, Rayna, Deborah Heath, and Karen-Sue Taussig. "Genealogical Dis-ease: Where Hereditary Abnormality, Biomedical Explanation, and Family Responsibility Meet." *Relative Values: Reconfiguring Kinship Studies*. Ed. Sarah Franklin and Susan McKinnon. Durham, NC: Duke University Press, 2001. 384-412.

Reardon, Jennifer. "Decoding Race and Human Difference in an Age of Genomics." *Differences: A Journal of Feminist Cultural Studies* 15.3 (2005): 38-65.

Richards, Martin. "Beware the Gene Genies." *Guardian* 14 February 2003. 10 May 2004 <www.guardian.co.uk/comment/story/0,,899835,00.html>.

Risch, Neil., Esteban Burchard, Elad Ziv, and Hua Tang. "Categorizations of Humans in Biomedical Research: Genes, Race, and Disease." *Genome Biology* 3.7 (2002): 2007.1-2007.12. 29 September 2003 <http://genomebiology.com/2002/3/7comment/2007>.

Rose, Nikolas and Carlos Novas. "Biological Citizenship." *Global Assemblages: Technology, Politics, and Ethics as Anthropological Problems.* Ed. Aihwa Ong and Stephen J. Collier. Malden, MA: Blackwell, 2005. 439-463.

Rose, Nikolas. *The Politics of Life Itself.* Princeton, NJ: Princeton University Press, 2006.

Rosenberg, N.A., et al. "Genetic Structure of Human Populations." *Science* 298 (2002): 2381-85.

Rosenthal, Carolyn J. "Kinkeeping in the Familial Division of Labor." *Journal of Marriage and the Family* 47.4 (1985): 965-74.

Rotimi, Charles. "Genetic Ancestry Tracing and the African Identity: A Double-Edged Sword?" *Developing World Bioethics* 3.2 (2003): 151-58.

Sarich, Vincent and Frank Miele. *Race: The Reality of Human Differences.* Boulder, CO: Westview Press, 2004.

Schneider, David M. *American Kinship: A Cultural Account.* Chicago, IL: University of Chicago Press, 1968.

Schwartz, Robert S. "Racial Profiling in Medical Research." *New England Journal of Medicine* 344.18 (2001): 1392-1393.

Skorecki, Karl et al. "Y Chromosomes of Jewish Priests." *Nature* 385 (1997): 32.

Smedley, Audrey. *Race in North America: Origin and Evolution of a Worldview.* Boulder, CO: Westview Press, 1998.

Wade, Nicholas. "In the Body of an Accounting Professor, A Little Bit of the Mongol Hordes." *The New York Times.* 6 June 2006. 2 April 2007 <www.nytimes.com/2006/06/06/science/06genghis.html?ex=1307246400&en=6e088f7b72599bf1&ei=5088>.

Wailoo, Keith. *Drawing Blood: Technology and Disease Identity in Twentieth-Century America.* Baltimore, MD: The Johns Hopkins University Press, 1997.

Wailoo, Keith and Stephen Pemberton. *The Troubled Dream of Genetic Medicine: Ethnicity and Innovation in Tay-Sachs, Cystic Fibrosis, and Sickle Cell Disease.* Baltimore, MD: The Johns Hopkins University Press, 2006.

Waters, Mary. *Black Identities: West Indian Immigrant Dreams and American Realities.* Cambridge, MA: Harvard University Press, 2001.

---. *Ethnic Options: Choosing Identities in America.* Berkeley, CA: University of California Press, 1990.

White House. "White House Remarks on Decoding of Genome." *New York Times* 27 June 2000: F8.

White House Press Secretary. *Text of the Remarks on the Completion of the First Survey of the Entire Human Genome Project*. 26 June 2000. 1 October 2003 <http://clinton5.nara.gov/WH/New/html/genome-20000626.html>.

"Yorkshire Clan Linked to Africa." British Broadcasting Corporation (BBC). 24 January 2007. 2 April 2007 <http://news.bbc.co.uk/2/hi/science/nature/6293333.stm>.

Sustainability and the Challenges of Race, Gender, and Poverty to Contemporary Scientific Cultures

Robin Morris Collin

The problems that we have created as the result of our industrial development policies threaten the fundamental systems on which all life depends. This is the contemporary challenge of sustainability. The search for answers to these problems challenges the limits of traditional scientific and technological thinking. In order to serve a contemporary role as problem solver for these challenges, scientific and technological cultures must evolve their methodologies and ethical cultures to include values designed to engage policy decisions. This is most necessary where science is working at the limits of its knowledge base and expertise. Sustainability challenges cultures of science and technology to transcend their traditional disciplinary limits. This means incorporating qualitative factors often excluded by scientific models, and engaging nonscientific influences such as race, age, gender, class, and world view. Today's solutions must inescapably encounter unresolved conflicts of the past including discredited values underlying colonialism, racism, and gender subordination. Two models of partnership between scientific and technological expertise for policy and legal judgments will be compared: the environmental justice models of public involvement and community-based science, and the evolving discipline of risk perception.

Introduction: The Contemporary Challenge of Sustainability

Scientific and technological innovation, coupled with industrialism and the politic of Empire, created great wealth in western societies. Wealth creation and development were enhanced by law and policies protecting private property, and encouraging the use of science and technology to establish industrial models of business. Some problems were solved in these cultures, including food security, clean fresh water, and control of basic childhood diseases. But over time, these same industrial development policies have fostered a next generation of problems which threaten to overwhelm the fundamental life systems on which all life depends. These are the challenges to sustainability. For example, these challenges include climate change, species extinctions, and drug resistant viruses. The Millenium Ecosystem Assessment[1] has reported on the contemporary state these human generated crises; crises which are the direct result of industrial development policy. The Ecosystem Assessment concluded that while scientific and technologically induced changes to the various systems within our environment have contributed to substantial net gains in human well-being and economic development, these gains have been achieved at growing costs. These costs occur in the form of the degradation of many ecosystems, and the exacerbation of poverty for some groups of people. It concludes that unless addressed, these problems will substantially diminish the benefits that future generations obtain from ecosystems.

This is the contemporary challenge of sustainability. These challenges are forcing cultures enriched by industrial development models to confront the limits of prosperity and dominance. The role of scientific and technological cultures in meeting these challenges has brought science and technology to an appreciation of their disciplinary limits.

I. The Role of Science and Technology in Meeting the Challenge of Sustainability: The Problem of Social Equity and "Eigen"-Values

a. Science, Technology, and Empire: Internalized, Shared Values

Historically, science and technology in the west shared a period of unprecedented growth with the growth of Empire. As the empires of Europe extended their control over the people and natural resources of ethnically distinct lands and territories, science and technology developed increasingly intensive applications for labor and resources in the production and consumption of material goods. These applications of scale consuming both labor (slavery) and natural resources (colonization) at previously unknown rates created vast wealth. This wealth together with its need for labor, land, and resources were defended by the military dominance of Empire.

[1] The Millennium Ecosystem Assessment (MA) was called for by United Nations Secretary-General Kofi Annan in 2000 in a report to the General Assembly entitled, "We the Peoples: The Role of the United Nations in the 21st Century" (Overview of the Millenium Ecosystem Assessment). The MA was initiated in 2001 with an objective to "assess the consequences of ecosystem change for human well-being and the scientific basis for actions needed to enhance the conservation and sustainable use of those systems and their contribution to human well-being." More than 1,360 experts worldwide have contributed to the project. Each part of the assessment has been scrutinized by governments, independent scientists, and other experts to ensure the robustness of its findings.

Science, technology and Empire collaborated to form the engine of development that produce "first world" (and third world) living conditions in our post colonial era. The success of the industrial model using intensive use of labor and natural resources for private gain embedded key values and paradigms in developed cultures.[2] These values tend to intuitively govern decision making and conduct, especially in times of uncertainty about facts or causation (Functowitz et al. 141-143). Without reflection, these values and paradigms continue to organize and shape the policies of developed countries, science and technology without any need for express instructions or agreement (Lélé 354). In order to meet the challenges of sustainability, the policies of developed countries, science and technology must consciously re-examine their implicitly held values and paradigms. The ability to examine and choose values and paradigms is what makes humans able to change, in this case to change in favor of sustainability. This kind of self examination and exercise of professional autonomy belongs to education and intentional ethical culture.

Science and technology fueled the growth of industrial development, especially its dependence on cheap natural resources, insatiable need for cheap labor including slaves, and generation of waste and pollution. For example, the Cotton Gin permitted the industrial scale processing of cotton fiber. To feed the machine, more cotton was planted and harvested in the American South. This caused the transfer of millions of slaves from other forms of agriculture and small farms, to large factory-style plantations in the South (Berlin 147). The conversion of fields to growing cotton impoverished the soil creating the need for synthetic fertilizers. Cotton was prone to insect infestation prompting the use of pesticides. The slurry of chemicals that run off these fields into the Mississippi River has changed the Delta into a toxic soup (Helvarg 152-153).

In the US, an extract-consume-waste based economy maintained profit by consistently undervaluing natural resources, and externalizing waste onto politically impotent groups. The material relationships between scientific cultures and empire were protected and solidified by legal constructs of private property, patent, and copyright, which created vast private wealth in patented technologies, and human chattel property, slaves, for much of its history. The results in contemporary terms are a wealth gap unequaled in other developed nations, and sinks of pollution and waste causing our ecosystems to deteriorate under the combined pressures of pollution, waste and neglect. Science and technology served as twin engines of devel-

[2] Paradigms are a powerful collection of internalized values and instructions that shape and direct behavior, often without the need for express instructions or agreements. These self executing instructions will have the effect of producing the same types of results or structures over and over again. What makes humans extraordinary is our ability to change our paradigms, create new ones, and our ability to choose among them.

opment, and their most important post colonial contribution will be to solve some of the problems it helped to create (Functowitz et al.141-143).

Science and technology have had an impressive interaction with empire. It is not mere coincidence that the Scientific Revolution accompanied the beginnings of western colonialism. History suggests that science, technology and Empire established a powerful relationship continuing to this day. That relationship undergirds powerful paradigm about development: Nature is endlessly resilient, science is the only source of true knowledge, and Technology can remediate any unproductive or unpleasant condition in nature. Even though we know that none of the above is true, they have influenced the economic policy choices we continue to make. The complexity around each of these truisms unravels all optimism founded upon them. We now recognize that nature is a complex web of interrelated systems each of which have no human technological equivalent (Hawken et al. 152-156). These webs of life on which all life depends provide "services" for which we have no equivalent. And they are failing under the weight of waste, pollution, and (some would add) over population (Nelson 74-76). But the most enduring and troubling consequence of these paradigms in science and technological development are the internalized values.

Science shares fundamental values with commerce (Jacobs 45). Science and technology have often allowed patrons to shape their inquiries allowing research and development to be directed by the needs of profit-based enterprises and warfare. Under this direction, the interests and needs of women, children, the poor, and the environment itself were irrelevant. These groups and their needs are still largely excluded from safe dosage experiments (cf. Kuehn), and community concerns about harm to the environment and human health are confounded by the null hypothesis (McGarvey et al. 100). Finally, the needs of the poor are largely invisible to the industrialized world view. This has meant blindness in science and technology as to the equity consequences of research and development.

Patronage has taken science and technology out of a direct relationship with the public good. Instead science and technology rely on a proxy relationship to the common good based upon assumption that business profitability is a proxy for social good. To meet the challenges of sustainability, leadership in science and technology will have to consciously engage in ethical self restraint. That culture of self restraint will have to deliberately confront the internalized values and assumptions that have shaped their professional behaviors. Without that self examination and choice, those same values will reassert themselves without the need for explicit agreement, or verbal discourse. These values become the "Eigenvalues" of science and technology and will undermine the ability of these cultures to transcend their roots (cf. Berressem in this volume). As Albert Einstein said, "The world we have made as a result of the level of thinking we have done thus far creates problems that we cannot solve at the same level at which we created them."

II. Science in the Middle: Sustainability Revives Unresolved Conflicts Of The Past: US Perspective

Solutions to the challenges of contemporary sustainability must engage the policies and values of the colonial and industrial age. Science and technology cannot engage contemporary policy choices without acknowledging historical policy choices like the choice of petroleum and coal as energy resources, the choice of slave labor, and the choice of imperial colonialism to propel western economic development. Many conflicts remain unresolved about the values of that model of development. For example, the ideological values underlying colonialism, racism, and gender subordination. Contemporary policy debates place science and technology in the middle of developing policy disputes, unfairly diverting critical attention from political and industrial decision makers. Nevertheless, once in the middle, attention turns to science and technology for answers. Based upon their traditional thinking they founder on disciplinary limitations. Public distrust and anger follow the inability of science and technology to deliver clear, unequivocal answers for policy choices that must be made.

a. Challenges To The Limits of The Traditional Role of Science And Scientists

Jane Lubchenco, President of the American Association for the Advancement of Science (AAAS) proposed a survey of present and former board members and past presidents to sample their ideas about the organization's mission. The board sent out a letter in October 1996 posing four questions: What are the major issues facing society? What is the role of science in addressing these issues? What are the major issues and challenges confronting science? What should the role of AAAS be? The answers they received are eloquent in identifying the challenges that contemporary problems of sustainability pose to the limits of traditional scientific thinking and analysis. There was nearly universal consensus about the scientific and technological issues facing contemporary human society:

- environmental change and degradation;
- population;
- public health, particularly emergent and re-emergent diseases;
- public food and energy; education;
- equity, including the global maldistribution of wealth; and
- The public's understanding of science and technology.

Another thread joining many responses was the consensus that in every area of concern, the traditional way that science had framed its mission was inadequate. The Report concludes:

> [W]e were struck by one common characteristic of the issues our colleagues had identified as most urgent. Each one, from population and the environment to the public's understanding of science, seemed to have radically outgrown its previously accepted conceptual framing. For each of these issues, new theories, explanations, and cause-effect relationships were appearing on the horizon. These paradigm shifts call for more creative forms of collaboration between scientists and society and for a broader range of disciplines and competencies to take part in the process. (Jasanoff et al. 2067)

The post modern model of a scientific culture must become a partnership between science and acknowledged socially constructed values. This partnership could engage policy decisions where science is working at the limits of its knowledge base and expertise. This means a model incorporating qualitative factors often excluded by scientific models, and engaging nonscientific influences such as race, age, gender, class, and world view. One example, the evolving discipline of risk perception shows how science can begin to engage a variety of previously excluded stakeholder interests. In addition, there are many new models of community based public participation in which science is one of many stakeholders represented in dialogues about local challenges to sustainability. These are described briefly below.

III. The Evolving Science Of Risk Assessment: Risk Perception, Management and Assessment

Risk assessment is a scientific tool to identify, quantify, and characterize exposure to various types of dangers. While danger is real, risk is a socially constructed approach to managing the impacts of danger, and social construction implies organizing values (McGregor). Risk as a social construction, relates to issues of management, and ultimately political power to make decisions in the face of uncertainty. This tool is used widely as an investment tool in private investing, insurance, international development funding, setting governmental budgeting funding priorities, determining regulatory agendas and setting policy in areas of attenuated and severe risk like climate change. Risk management makes decisions concerning safe exposures, including acceptable risks of loss or harms. Risk management is also actively engaged in communication of these safety judgments. Risk perception explores how the social construction of risk is affected by voluntariness, world view, race, gender, education and other social factors (Finucane 159).

IV. Public Participation Models: Science as One of Many Stakeholders

Contemporary models of public involvement such as the Aarhus Convention, the Environmental Justice Model Plan for Public Involvement, and the Environmental Justice Collaborative Guide, recognize the critical role that openness, transparency, accountability, and public involvement play in order for a democracy to meet the challenges of sustainability. Process is how we make necessary decisions in a democratic society faced with indeterminacy and uncertainty. Democratic process creates the essential link between concept and community; democracy ensures that the link is legitimate. Democratic decision making implies a role for lay communities in all legitimate decision making; exclusion and problems of exclusion undermine the fundamental legitimacy of decisions made in a democracy. The democratic premise in the US, and also in the worldwide Civil Society movement, is that non-governmental organizations (including grass roots community organizations) play a vital role in the shaping and implementation of participatory democracy. Specifically, an engaged and well informed citizenry can better participate in decision making, build support for decisions, and strengthen their implementation. Community participation expands and improves all stages of development from the identification of problems, setting of priorities, to developing just and effective solutions.

All these models share some common characteristics which augment and extend the policy discourse associated with challenges to sustainability beyond scientific expertise. They acknowledge community knowledge about place, specifically knowing the people, flora, fauna, culture and history of the place, leading to the use of the term Place Study rather than Case Study. The community is also involved in problem identification and definition which helps to include aspects such as race, age, gender, class, which might have been excluded under a traditional scientific protocol. These models also provide a place for community knowledge and involvement thus moving towards Place-based ecology and sustainability.

Conclusions

In the search for solutions to the challenges of sustainability there is a role for community based and others along with science and technology. Science and technology alone, operating within their usual constraints, simply cannot provide answers that will change the trajectory our ecosystems, environments and human communities are on. Many scientists have urged science and technology to search for and find a new method that will open up new dimensions. Usable uncertainty, the intentional incorporation of positive values, and new peer relationships are all part of science and technology that can meet the challenges of sustainability (Functowitz et al. 141-143).

Works Cited

Berlin, Ira. *Generations of Captivity: A History of African-American Slaves*. Cambridge, MA: Belknap Press of Harvard University Press, 2003.

Berressem, Hanjo. "*The Habit of Saying I*: Eigenvalues and Autopoeisis." (article in this volume).

Convention on "Access to Information, Public Participation in Decision-Making and Access to Justice in Environmental Matters." 25 June 1998 (entered into force 30 October 2001). *International Legal Materials* 38 (1998): 517-533. 27 October 2007 <www.unece.org/env/pp/treatytext.htm>.

Finucane, Melissa L., Paul Slovic, C.K. Mertz, James Flynn, and Theresa A. Satterfield. "Gender, Race, and Perceived Risk: The 'White Male' Effect." *Health, Risk and Society* 2 (2000): 159-172.

Flynn, James, Paul Slovic, and C.K Mertz. "Gender, Race, and Perception of Environmental Health Risks." *Risk Analysis: An International Journal* 14 (1994): 1101-1108.

Functowitz, Silvio O. and Jerome R. Ravetz. "A New Scientific Methodology for Global Environmental Issues." *Ecological Economics: The Science and Management of Sustainability*. Ed. Robert Costanza. New York: Columbia University Press, 1999. 137-151.

Hawken, Paul, Amory Lovins, and L. Hunter Lovins. *Natural Capitalism*. Boston, MA: Little, Brown and Co., 1999.

Helvarg, David. *Blue Frontier: Saving America's Living Seas*. New York: W.H. Freeman, 2001.

Jacobs, Jane. *Systems of Survival*. New York: Random House, 1992.

Jasanoff, Sheila, Rita Colwell, Mildred S. Dresselhaus, Robert D. Goldman, M.R.C. Greenwood, Alice S. Huang, William Lester, Simon A. Levin, Marcia C. Linn, Jane Lubchenco, Michael J. Novacek, Anna C. Roosevelt, Jean E. Taylor, and Nancy Wexler. "Conversations with the Community: AAAS at the Millenium." *Science* 19 December 1997: 2066-2067.

Kuehn, Robert R. "The Environmental Justice Implications of Quantitative Risk Assessment." *Environmental Justice: Law, Policy, and Regulation*. Clifford Rechtschaffen & Eileen Gauna. Durham, NC: Carolina Academic Press, 2002.

Lélé, Sharachchandra and Richard B. Norgaard. "Sustainability and the Scientist's Burden." *Conservation Biology* 10 (1996): 354-365.

McGarvey, Daniel J. and Brett Marshall. “Making Senses of Scientists and ‘Sound Science’: Truth and Consequences for Endangered Species in the Klamath Basin and Beyond.” *Ecology Law Quarterly* 32 (2005): 73-111.

McGregor, Donald G., Paul Slovic, and Torbjörn Malmfors. “‘How Exposed is Exposed Enough?’ Lay Inferences About Chemical Exposure.” *Risk Analysis* 19 (1999): 649-659.

Nelson, Gerald C. “Drivers of Ecosystem Change: Summary Chapter.” 12 Dec. 2005. Millenium Ecosystems Assessment Report, Ecosystems and Human Well-Being: Current State and Trends. 27 October 2007 <www. http://www.maweb.org/documents/document.272.aspx.pdf>.

Overview of the Millenium Ecosystem Assessment. 27 October 2007 <http://www.maweb.org/en/About.aspx#1>.

United States. Environmental Protection Agency. 2002 Feb. “Environmental Justice Collaborative Model: A Framework to Ensure Local Problem-Solving.” EPA-300-R-02-001. 27 October 2007 <http://www.epa.gov/compliance/resources/publications/ej/interagency/iwg-status-02042002.pdf>.

United States. National Environmental Justice Advisory Council. 2000 Feb. “The Model Plan for Public Involvement.” EPA-300-K-00-00. 27 October 2007 <http://www.epa.gov/compliance/resources/publications/ej/model_public_part_plan.pdf>.

Constructing Competitive Advantage: The Evolution of State R&D Investment Funds in the United States

Heike Mayer

Introduction

With the rise of the knowledge-based economy, policymakers at the state and local level in the United States have become increasingly aware of the importance of research and development (R&D) activities to economic growth and development. State investments in university research and innovation are beginning to fill a void left by decreasing levels of federal funding (see Figure 1). While states have a long tradition in economic development—primarily through the attraction and retention of firms—policymakers are beginning to reorient their programs thereby making them more strategic and focused on supporting specific technologies and industry sectors. Parallel to the evolution of state R&D investments, industry sectors such as high-technology and biotechnology have significantly changed the ways in which innovation is organized at the intra-firm level. This involved a move away from conducting research and development in in-house corporate R&D laboratories (such as Bell Labs or Xerox PARC) to a more open approach that is based on the assumption that innovative ideas come from many different sources including consumers, customers, suppliers, and most importantly universities (cf. Chesbrough). Some states are realizing the opportunities this new model offers and are providing funding for new kinds of university-industry partnerships. Increasingly, states have become pro-active R&D investors and policymakers often assume that they will be rewarded with significant economic growth and development.

Examining public policies will shed light on the way policymakers and the public perceive science and technology and its use in the public realm. The discursive practices focus on the use of science for economic development benefits. In other words, innovation conducted at local universities and firms and funded by state governments is expected to materialize through the creation of firms and jobs for local residents and through income gained from commercialization and technology transfer. Such a perspective, however, is very limited and presumes that innovation—the application of ideas and inventions—follows the so-called linear or science-push model. The linear model has been widely criticized for not taking into account that the way industry approaches innovation has changed and that firms are keenly aware of the wide range of sources for innovative ideas (Cooke and Morgan 12). Innovation in essence is what Freeman and Soete call a "two-sided or coupling activity" that involves supply (technology, ideas, innovations) and demand (market, consumer, user) (Freeman and Soete 200). While many programs still focus on the linear model, some state programs evolved and seem to build on an interactive model of innovation rather than the linear model. To be effective and to follow through on the political promises for economic prosperity, state R&D investment programs need to take the interactive innovation model into account. Cooke and Morgan argue that this model "carries radical implications not just for firms but for a wide array of public and private institutions. In particular, the interactive character of the innovation process means that to be effective, firms, regions, and nations need to develop organizational structures and mechanisms which promote continuous interaction and feedback within and between firms and among the various institutions which constitute the national system of innovation" (13).

This chapter will review the evolution of state R&D investments funds. The research is the result of a project sponsored by the National Governors Association and the Pew Center on the States aimed at making recommendations to Governors and their staff about how to effectively invest in research and development (Pew Center on the States and National Governors Association). Over the last decade, states have increasingly made available dedicated funding for research and development in certain science and technology fields. All kinds of states, large and small, are focusing on this kind of economic development policy. As would be expected, California is at the forefront with its investments in stem cell research, microelectronic and computer science among other fields. Increasingly, however, small states (for example West Virginia, South and North Dakota, etc.) and states trying to step out of California's or Massachusetts's shadows are making bets through innovation investments. The chapter will illustrate how Oregon, Idaho, Arizona, and Kansas are utilizing state investments to advance their economic fortunes. A critical analysis of the intentions and promises as well as failures will be presented.

Evolution of State R&D Investment Funds

State investments in research and development evolved in very distinct ways over the last two decades. During the 1980s many states invested in building capacity at their research universities. Only a few programs focused on university-industry partnerships. During the 1990s, there seemed to have been a pause in initiatives. During this period states with significant need to create research capacity such as Missouri and Kentucky made major investments in their higher education infrastructure. Beginning in 2000, however, many other states became active and adopted a more strategic focus both in terms of program development and science areas benefiting from funding. Many began to strategically target their investments on certain science or technology areas (i.e. engineering, computer sciences, biosciences, stem cells etc.). In addition, state leaders became more creative regarding university-industry and/or university-university partnerships. This evolution may indicate that states have become more sophisticated regarding their R&D investments and that their funds help universities play a critical role in the emerging interactive and open innovation models.

1980s: Building Research Capacity

As Table 1 outlines, most state efforts to invest in R&D that started in the 1980s were aimed at building research capacity through funding endowed chair or professorship positions. States also invested in the establishment of university-based centers of excellence. The primary goal of these programs was to strengthen the research capacity and to create areas of expertise. Generally, programs were not targeted towards certain technologies or research fields. In Tennessee, for example, centers of excellence created during this time focus on a wide range of fields such as popular music, egyptology, and neurosciences. Sometimes programs were created through wide-ranging education reforms that not only touched the institutions of higher education but also the K-12 and community college systems (i.e. Tennessee and Kentucky). These reforms mostly took place in states that needed to build their educational infrastructure and capacity. By now, funds for endowed faculty positions have dried up (i.e. Kentucky, Missouri). Other states have successfully transitioned into a new phase of investments (i.e. Tennessee).

Few programs that were started during this period aimed at building university-industry partnerships during the 1980s. Maryland's Industrial Partnerships Program (MIPS) and Connecticut's Yankee Ingenuity Technology Program are examples of pioneering initiatives focused on connecting university and industry. Maryland's program has been very successful in supporting research that yielded significant economic benefit: For example, funding was made available for research involving MedImmune's best-selling drug Synagis (Maryland Industrial Partnerships).

2000s: Constructing Competitive Advantage

Beginning in 2000, states started to strategically focus their R&D investments. The major goal was to construct competitive advantage by focusing on certain technology areas, leveraging federal funding and connecting universities with industry. The efforts coincided with a decrease in federal funding and a heightened awareness among states about the competitive nature of federal investments in R&D.

Massachusetts stands out for its efforts to leverage federal and industry funding. The state's Research Center Matching Fund program supports university efforts aimed at attracting federal funding for university-industry centers. Virginia's Commonwealth Technology Fund supports collaborative research between universities and industry partners. This fund, however, has been plagued by budgetary problems—a common occurrence and barrier to success of this type of public policy—and by a strong university lobby to appropriate money directly to their institutions. Tennessee is banking on the partnership between the University of Tennessee and Oak Ridge National Laboratory.

Popular among states is to focus investments on life sciences research. Biotechnology or life science in general has become a desirable industry for state and local economic developers to pursue (Cortright and Mayer). Kansas, for example, has made major investments in the biosciences as will be discussed later on. States with stem cell research funds represent a special case: In response to President Bush's 2001 veto of stem cell research funding, Connecticut and Maryland created funds to support this kind of research. Their funding ranges between $15 million and $100 million and pales in comparison to California's $3 billion bond-financed program. Maryland developed a Stem Cell Research Program that explicitly focuses on translational research and funds research that would otherwise not be funded by federal agencies such as the National Institutes of Health would the veto not exist. The issue of stem cell research, however, is not everywhere blissful: In Missouri, opposition, which is primarily motivated by ideological and religious concerns, to stem cell research—in addition to budgetary problems—has stymied the state's efforts to invest in higher education. In smaller states—such as West Virginia and Idaho—proposals to support higher education are discussed at the time this chapter was written. States where urban-rural differences are pronounced face more difficulties in gaining support for their investments because rural constituencies typically have a hard time understanding the value of science and technology investments. One example is Idaho which will be discussed in depth later in this chapter.

In many states, R&D investment funds are managed by public-private agencies or by universities. Agencies such as Connecticut Innovations, the Massachusetts Technology Collaborative, or Virginia's Center for Innovative Technology benefit from their experience with other types of funds and their ties to industry. Such public-private agencies are often deemed to be more flexible and dynamic (albeit less accountable) compared to state government offices or departments. It is important to note that state R&D investment funds do not exist in a vacuum. Most states offer a portfolio of funds focused on research and development, commercialization, technology transfer, and business development.

Table 1: Evolution of Select State R&D Funds

Time	Program Type	States
1980s	Chairs & Centers of Excellence Programs	KS: Centers of Excellence (1980s) VA: Technology Development Centers (1980s) TN: Chairs of Excellence & Centers of Excellence (1984) CT: Endowed Chair Investment Fund (1985) MO: Food for the 21st Century (1985) MD: Biotechnology Institute (1985)
	University-Industry Partnership Programs	CT: Yankee Ingenuity Technology Program (1985) MD: Maryland Industrial Partnerships Program (1987)
1990s	Chair Programs	MO: State-Endowed Chair & Professorship Program (1995) KY: Bucks for Brains (1997)
2000s	University-Industry and/or University-University Partnerships	VA: Commonwealth Technology Research Fund (2000) MA: Research Center Matching Funds (2003) TN: Governor's Chair & Joint Institutes (2004)
	Leveraging federal funding	KY: KSEF R&D Excellence Program (2000) WV: Research Challenge Trust Fund (2003)
	Life Sciences / Biotech	KS: University Research and Development Act (2001) MO: Life Sciences Research Trust Fund (2003) KS: Kansas Bioscience Initiative (2004) KS: Kansas Cancer Center (2006)
	Stem Cell Funds	CT: Stem Cell Research Fund (2005) MD: Stem Cell Research Program (2006)
	Major HEI investments	VA: Governor's Higher Education Research Initiative (2005)
	Proposals	WV: Vision 2015 (2006) ID: S&T Council Proposal & Governor's Proposal (2006) MO: MOHELA sale (2006/2007)

Note: Table does not include an analysis are state venture capital and commercialization funds.
Source: (HeikeMayer. *A Review of State R&D Investment Funds: Ten Case Studies*)

Why do states invest in research and development?

Traditionally, states have focused their economic development efforts on recruiting and attracting facilities and firms. This type of economic development policy is commonly referred to as *smokestack chasing*. The practice originated in Southern states such as Mississipi and Alabama in the 1970s as a result of declines in manufacturing employment and plant closings. Starting in the 1980s, however, state governments began to also focus on promoting technology-based economic development, innovation, entrepreneurship, and workforce development. While they did not abandon the practice of smokestack chasing, they added new programs and strategies such as R&D investment funds, state-led venture capital funds, commercialization and technology transfer programs to their policy repertoire (Clarke and Gaile). Many state leaders and policymakers were motivated by the experiences of Silicon Valley, Boston's Route 128, and North Carolina's Research Triangle Park and began to see universities and other higher education institutions as critical actors in the quest to make their economies more competitive and resilient. In contrast to smokestack chasing practices, the focus of policies and programs shifted towards investments that would not only support the building of facilities or infrastructure—so-called bricks and mortar projects—but also investments in talent, research, entrepreneurship and innovation (Plosila).

During the 1980s and 1990s, most states began to invest in R&D activities. The main focus is on supporting science and technology to the benefit of improving the state's economy. In some states, such investments are made through public-private intermediaries that manage these funds. Funding is available for building and expanding facilities (such as university laboratories, dedicated research centers, incubators, etc.), collaboration between university and industry and between different universities in the state or across state borders and support for research. Investments in the latter are made with the hope that direct support for research would leverage additional funding from federal sources such as the National Science Foundation or the National Institutes of Health. Most often investments are financed from the state's general fund, but sometimes unique financing tools such as bonds, Tobacco settlement money, earmarked taxes or tax increment revenues are used (National Governors Association and Pew Center on the States).

Often a fairly simplistic thinking accompanies such investments: For example, Nevada's Governor stated in his 2007 State of the State Address that his state will "hire world-class researchers who will bring their work to Nevada. This will provide high-end jobs for Nevadans" (cf. Gibbons). His statement echoes the by-gone era of trying to recruit firms. Instead of trying to chase factories with smokestacks, Nevada, it seems from the Governor's statement, is trying to chase scientists. Kentucky made its intentions very clear and aptly called its investments the "Bucks for Brains" program and since its inception in 1997 has been able to increase the number of endowed chairs from 55 to 212 and professorships from 53 to 312 from 1997 to 2003.[1]

[1] See Mayer, *A Review of State R&D Investment Funds: Ten Case Studies.*

Various factors such as the rise of the knowledge-based economy, the potential of developing innovative regions modeled after Silicon Valley, substantive economic restructuring and industrial decline, the globalization of R&D, and the recognition that federal funding for research is declining while industry is beginning to search for innovative ideas developed outside its corporate laboratories motivate governors and legislators. While it is often argued that the United States do not engage in industrial policy (when compared to other countries such as Japan or Korea for example), these R&D investment funds can be considered *de facto* industrial policy at the level of the state. Most states are biased towards supporting industries such as high-technology and biotechnology and, motivated by the discussion of global climate change and its potential for economic development, increasingly they focus on alternative energy and fuels technology.

In the transatlantic context, the experience of individual states in the US is neither new nor unprecedented. The case of Germany, for example, may point to a similar development. Funding for higher education is provided by the states (Länder) and the federal government, with research investments coming primarily from the latter. Some German states, however, have begun to invest in innovation and entrepreneurship towards the advancement of certain regional economies. During the 1990s, the "Bavarian government started initiatives that were aimed at upgrading the research infrastructure and the provision of risk capital at the regional level" (Kaiser 847). These state investments enabled Munich to become one of Germany's premier high-technology regions (cf. Sternberg and Tamasy). Finland is another interesting example. With a population of more than 5 million, Finland is as large as Maryland, Arizona, Tennessee, or Wisconsin. Finland's science and technology policy go back to the 1960s and 1970s and they stand out because the emphasis was on "technical research, technical faculties, research institutes and firms, instead of more science and university-based policy" (Oinas 1234). In addition, the country adopted regionalized approaches to innovation and its programs place a heavy emphasis on cooperation between firms, universities, research institutions and even foreign partners, creating an environment in which interactive or open innovation can flourish.

Case Studies: Boosting research to catch up

The case studies presented here illustrate how states that have traditionally lagged behind in terms of economic development and prosperity are catching up. I will discuss the experiences in Oregon, Idaho, Kansas, and Arizona. These four states can generally be considered second tier or emerging when it comes to policy innovations in research and development. They generally lack behind pioneering states such as California, New York or Massachusetts. A common theme among the four states discussed here is that each state is experiencing tremendous shifts in the types of industries that are driving their economies. Oregon and Idaho, for example, have transitioned from being reliant on resource-based sectors such as timber and agriculture to high-technology manufacturing (primarily semiconductor manufacturing). Kansas is a state that has traditionally served as an important logistics location in the Midwest (railroad) and its economy was heavily based on agriculture. Nowadays Kansas' major urban centers like Kansas City and Wichita are reliant on advanced manufacturing, life science and information technology (Mayer, "Completing the Puzzle: Creating a High Tech and Life Science Economy in Kansas City"). Arizona represents a state that is making a transition from an economy focused on amenities and resources—typically characterized by the so-called five Cs: climate (tourism and retirees), cotton, citrus, copper (mining), and cattle—to one aimed at utilizing research institutions and universities as economic drivers. In addition, each state illustrates interesting cultures and political economies that challenge the implementation of public policies aimed at transitioning into knowledge-based economies.

Oregon

Oregon, a state of about 3 million inhabitants that borders Washington and California in the Pacific Northwest, traditionally focused on resource-based industry sectors such as timber, wood products and agriculture. Lesser known, however, is the state's history and strength in high-technology manufacturing. Since the mid-1940s, a small but innovative high-technology sector formed in its main population center Portland. The industry is characterized by a concentration of manufacturers of electronic test and measurement instruments, semiconductors, specialized computers and semiconductor manufacturing equipment producers, and software publishers. Like many other high-technology regions, Portland has a nickname and the concentration of firms is often referred to as the *Silicon Forest* (Mayer, "Taking Root in the Silicon Forest: The Role of High Technology Firms as Surrogate Universities in Portland, Oregon").

Interestingly the region emerged as a high-tech location in the absence of a major research university such as MIT or Stanford. Instead it utilized the presence of major high-technology employers such as Tektronix and Intel. These firms have been able to attract, retain and develop a qualified labor pool. They are conducting innovative research and development activities that sometimes spill over into the region in the form of startup companies or talent that moves from one company to another. And lastly, the firms have functioned as incubators for more than 150 spinoff companies. In sum, Tektronix and Intel functioned as so-called "surrogate universities" and helped built a specialized high-technology agglomeration (Mayer,

"Taking Root"). Along with serving the region as "surrogate universities," Tektronix and Intel have exerted significant influence on policymaking especially with regard to improving higher education offerings and infrastructure (Mayer "Competition for High-Tech Jobs in Second Tier Regions: The Case of Portland, Oregon"). These major corporations along with other high-technology firms, interest groups and individuals such as venture capitalists, university representatives, and local service providers (specialized lawyers etc.) have formed a strong lobby to influence the state and push for increased investments in higher education and research. They were motivated by the fear to loose the region's competitiveness because of the lack of an appropriate research and development infrastructure. In addition, firms like Tektronix and Intel are less interested in conducting research and development in-house. These firms adopt the new open innovation paradigm and they focus on the benefits of external sources.

Lobbying efforts to improve higher education started in the late 1950s when the Governor at the time (Mark Hatfield) called for a report on science and engineering education and subsequently appointed an industry-led advisory committee in 1959 (cf. Dodds and Wollner). As a result of these early collaborations, the region established a graduate-only engineering and science institute modeled after the Massachusetts Institute of Technology (MIT). Due to funding constraints, the Oregon Graduate Institute—which is now part of Oregon Health and Science University—never achieved a standing similar to its famous East Coast model. Over the years, industry leaders kept complaining about the lack of an appropriate higher education infrastructure. While early recommendations and suggestions primarily focused on improving the workforce and education offerings, initiatives in the 1990s and 2000s began to emphasize improving the R&D infrastructure. In addition, an explicit argument has been made that these public investments in R&D would improve the state's and Portland's economic competitiveness. From 2001 to 2005 a Governor appointed committee deliberated about the best ways in which Oregon could improve its support for innovation, research and development. The resulting strategy built the foundation for a more focused approach to invest in so-called "signature research" areas that built on the state's economic strength in certain industry clusters such as high-technology (Oregon Council for Knowledge and Economic Development). Industry leaders recommended that the most promising "signature research" area would be focused on nanotechnology. The state subsequently invested about $21 million into the creation of the so-called Oregon Nanoscience and Microtechnologies Institute (ONAMI), which started operation in 2003. The Institute reflects a very different approach to university-industry relationships and may come close to the interactive innovation model mentioned at the outset of this chapter. ONAMI is a collaborative effort between the state's three major public research universities, a federal research laboratory (Pacific Northwest National Laboratory) and more than 20 high-technology firms (so-called "research and commercialization partners"). Universities and industry both have a stake in the governance structure of the Institute. In addition, there are tight cooperative links between them through research programs, shared facilities, internship programs, and commercialization agreements. So far, ONAMI has been successful in leveraging federal funds (about $20 million) and commercializing technology through he creation of two companies (Oregon InC).

Oregon is continuing with its more strategic approach to focus on research and development and the most recent innovation strategy targets specific industries and technologies. In 2005, the Governor and the Legislature created the Oregon Innovation Council. The council is comprised of 40 business, higher education and government leaders and it created the Oregon Innovation Plan (Oregon InC). The 2002 report of the Oregon Council for Knowledge and Economic Development (Oregon Council for Knowledge and Economic Development), for example, acknowledged workforce development as one of three recommendations. In 2007, however, Oregon InC's innovation plan embeds workforce and skill development issues within the industry-based and research center initiatives. Industry-based initiatives focus on ocean wave energy, food processing/seafood and manufacturing. Signature research center support will go towards ONAMI and towards institute dedicated to biotechnology, infectious disease and drug development. In addition, several recommendations are made regarding a more supportive environment for innovation and entrepreneurship (i.e. angel and venture capital investments, innovation acceleration, technology transfer etc.).

The evolution of Oregon's R&D investment efforts illustrates a set of insights into the politics of economic development. Being a resource-based economy, Oregon's political culture has not valued higher education and therefore the state's policymakers have not invested to a great extent in their universities. Oregon's economy, however, has dramatically changed over the years. Knowledge-based industries are now driving its prosperity, evidenced by the high-technology industry's large share of the state's exports. Industry, in turn, has become the main advocate for a more pro-active and strategically focused approach to state R&D investments.

Idaho

Oregon's neighboring state Idaho has been less successful in making R&D investments a priority. The state's political culture and emphasis on traditional industries such as agriculture (dairy, potatoes, etc.) and a strong lobby on behalf of these traditional industries has curtailed investments in knowledge-based industries. While traditional sectors are still important to Idaho's economy, the high-technology industry has become a major economic driver. Idaho's capitol city Boise is home to firms such as Micron Technology, a semiconductor manufacturer, and Hewlett-Packard. The Silicon Valley-based Hewlett-Packard established a branch facility in Idaho in 1978 due to low cost and labor availability. Since then, Hewlett-Packard changed its Idaho operation from a manufacturing-based to an innovation-based facility and its engineers, software programmers and marketing experts have been responsible for all the inventions associated with the laser jet printer. Micron Technology is a home-grown company that was founded by a team of engineers who grew up in the state and moved back to Idaho after working for Mostek, a pioneering semiconductor company located in Dallas, Texas. The firm is one of a few US-based firms that still produce semiconductor memory chips within the United States. Most companies in this sector have outsourced their manufacturing to Asian countries. Similar to the Portland case, Micron and Hewlett-Packard contributed to Boise's emergence as a high-technology region. The region has become very innovative—as evidenced by high levels of patent registrations—and entrepreneurial with many former employees of Micron Technology and Hewlett-Packard starting and growing companies.

In contrast to Oregon, Micron and Hewlett-Packard have not been as pro-active in pushing the state to improve the higher education infrastructure. Both firms have significantly contributed through private donations to the creation of specialized education and research programs at local universities (primarily at Boise State University). During the late 1990s, state leaders, however, recognized the need to be more strategic regarding their efforts to support science and technology. In 1999, Governor Kempthorne announced the formation of the Idaho Science and Technology Advisory Council. The council developed the state's first strategic plan, which was released in 2001. The strategy recommended investments in science and technology education, R&D and university-industry collaborations. The recommendations were incorporated in the Governor's budget, but were not approved by the Legislature. Three years later, in 2004, the initial plan was updated and a new attempt at requesting funding for R&D investments was made. The only budget request legislatures funded at the time were rural initiatives such as a $5 million investment in rural broadband services, reflecting mostly rural interest of the state's legislatures. In 2006, the Council made a third attempt and proposed a $48.8 million economic stimulus package incorporating investments in education, research and development, marketing, business incubators, and technology infrastructure. The state's new Governor, a native of Idaho with close ties to the agricultural industries, did not take up the Council's recommendations in his budget request and instead recommended investments in energy studies, dairy and animal research.

Idaho is a state with a strong urban-rural divide. Many of the state's residents and lawmakers may not realize the significance of knowledge-based industries because they have traditionally been reliant on resource-based industries. Such a divide may explain the lack of innovation in state policymaking regarding R&D investments. Entrepreneurs and industry representatives, however, have become impatient. In early 2007, members of the Science and Technology Advisory Council expressed concerns and disappointment over the lack of support for their recommendations by the Governor (Kovsky). The Council seems to have introduced a new voice into the policymaking process in Idaho. Most of its members are representatives from high-technology firms or research institutions. Some are newcomers to the state and most operate their firms or conduct their business activities in a global context. These new voices reflect the significant demographic and economic changes that are taking place in the state. For example, Boise—Idaho's main population center and location of the high-tech industry—has only recently experienced tremendous urban growth and development. With this expansion comes the attraction of newcomers—in-migrants and immigrants—that bring a new outlook and perspective. Often these newcomers are from California, Washington and Oregon, states where policymakers have already recognized the importance of the new economy.

Kansas

The state of Kansas is situated in the American Midwest and is part of an area that is often referred to as the "American Heartland." Like Oregon and Idaho, the state's economy has traditionally thrived on agriculture. The state—and especially its main urban center, Kansas City—has played a pivotal role in moving goods from the rural hinterland to urban markets to the West and East. Kansas City was founded on the confluence of the Missouri and Kansas Rivers and became a trading hub for livestock, grain and other freight. Interestingly, the legacy of Kansas as a location for trading livestock has carried over into the new economy and is evidenced in the fact that the Greater Kansas City region is home to a large share of firms operating in the animal health and nutrition market. Companies like Böhringer Ingelheim, Hill's Pet Nutrition, Fort Dodge Animal Health and others are located in the region and account for "nearly 32 percent of total sales in the $15.2 billion global animal health market" (Animal Health Corridor). In addition, the region has a legacy in pharmaceutical industries, telecommunications and aerospace.

Recognizing these assets, state leaders began to invest in the bioscience industry in 2001. Then, the legislature passed the University Research and Development Act, which authorized about $130 million in bonds to finance various bioscience facilities at Kansas universities. These investments, however, represent a more strategic focus on one type of industry, namely bioscience. Since about the mid 1980s, Kansas supported Centers of Excellence at universities and their focus has been more broad-based than just a single industry. Centers were established to help advanced manufacturing, information and telecommunication, polymer research and aviation research. In addition, these centers were spread across universities thereby aligning with the geographic locus of the industry.

The focus on bioscience was spurred by the philanthropic efforts of a Kansas City-based entrepreneur in the late 90s and early 2000s. In 2000, Jim Stowers, together with his wife Virginia, donated more than $2 billion and created the Stowers Institute for Medical Research (see Figure 2). More than 300 scientists conduct basic research on genes and proteins. The establishment of the institute encouraged local leaders and policymakers to invest in the nearby higher education infrastructure. The Kansas City Area Life Science Institute (KCALSI) was formed with the goal to increase research expenditures at area research institutions and hospitals. The 2001 University Research and Development Act brought much needed investments in facilities. In 2004, Governor Sebelius signed into law the Kansas Economic Growth Act (KEGA), a $500 million initiative to invest in bioscience and entrepreneurship. KEGA established the Kansas Bioscience Authority (KBA) as an intermediary responsible for making investments in the industry. The Authority manages six programs that are primarily focused on increasing research at local universities and encouraging university-industry partnerships. The Authority also gives R&D incentives to companies conducting bioscience research in collaboration with universities or research institutions. Recently, the Authority with support from the Governor's office has focused its attention on the recruitment of the National Bio- and Agro-Defense Facility, a $450 million laboratory commissioned by the Department of Homeland Security. Kansas is among the finalist and competes with Georgia, North Carolina and Texas. A winner will be picked by October 2008 (Gertzen). These recruitment efforts look more like smokestack chasing than strategic investments in a new economy. A dedicated website was set up (http://www.nbafinkansas.org/), KBA made $250,000 available and the Governor appointed a 45-person task force to make sure "that this "once-in-a-lifetime opportunity" to build a federally-funded, state-of-the-art, 500,000 square foot National Bio and Agro-Defense Facility (NBAF) becomes a reality for Kansas" (Heartland Bio Agro Consortium).

Kansas neighboring state Missouri is less proactive in making state investments in research and development because political opposition has stalled the efforts. While Idaho and Oregon represent states where rural and agricultural interests have been strong and often in opposition, Missouri represents the conservative religious interests that are adamantly opposed to investments in life science and stem cell research. In 2006, for example, a ballot was introduced in Missouri that would allow stem cell research or treatment to occur in the state. This represented a direct response to the efforts of Missouri policymakers to introduce laws that would make such research a criminal offense (New York Times). Even though the ballot was heavily supported by the research and business community—the founders of the

Stowers Institute contributed $30 million to the campaign—the amendment passed only with a slim margin (51 percent to 49 percent) and the political environment did not change much (cf. Davey). Stowers, for example, suspended plans for a $300 million expansion and higher education investments are blocked by conservative legislators. The fundamentalist opposition to bioscience research may be seen as a "reaction to the process of modernization, liberalization and secularization of the public sphere" whereby simple solutions such as the criminalization of research on stem cells are sought (Ostendorf 17) and investments in the economy are boycotted.

Arizona

Among the states discussed here, Arizona represents the most far-reaching and visionary efforts in promoting science and technology. The state has adopted an aggressive leapfrog strategy aimed at creating a bioscience economy. It has made large investments in science and technology and is working on improving its higher education infrastructure. The state's population seems to recognize these efforts: A recent study reports that 59.5 percent of Arizonans agree that science and technology are important to the state's economic development (Morrison Institute for Public Policy).

Arizona's strategic efforts in economic development began in the early 1990s when the state started to develop a comprehensive plan to create a competitive global economy. The plan was developed through the participation of more than 1,000 Arizonans. As a result, industry working groups were formed (cf. Waits). Arizona utilized these "cluster working groups to help policy makers better understand an industry, the challenges it faces, and the most valuable assistance government can provide" (Waits 39). These working groups still exist and some are very active in networking, lobbying and policymaking. In particular the biotechnology cluster group, which formed in 1997, belonged to the most active groups and its members were instrumental in reorienting public policy towards the creation of a life science economy in Arizona.

During the late 1990s and early 2000s, Arizona leaders perceived their economy in decline and they recognized windows of opportunity to make significant changes and address economic restructuring. The terrorist attacks on September 11, 2001 contributed to a downturn in the tourism industry. In addition, the natural resource economy—especially copper mining—suffered and the real estate industry started to slow. While these structural changes took place, several opportunities emerged and coalesced: A gubernatorial change reoriented policy priorities towards public investments in education, a ballot initiative allowed the use of sales tax revenues for investments in higher education institutions, an opportunity to attract a biotechnology research organization to Phoenix emerged, and new leadership at Arizona State University reenergized efforts around building a life science economy. In addition, Arizona started to copy Ireland's efforts in investing in science and technology and founded the Science Foundation Arizona. Some of these initiatives represented a significant turn in the way Arizonans thought about the knowledge economy.

A significant turn in public policy related to education started to take place in Arizona during the early and mid 1990s. Until then, the state did not pay much attention on education issues. Education funding, for example, decreased during Governor Symington's six year term from 1991 to 1997 (cf. White). Symington resigned in 1997 and Interim Governor Jane Dee Hull was elected in 1998. The new Governor pledged to make education a priority and in 2000, she introduced a proposal to fund education through a 0.6 percent sales tax increase. The proposal was strongly supported by the business community partially because the initiative not only included K-12 funding but also significant investments in universities. "A coalition of the business community, the education community, the Governor's office, and the Superintendent conducted "the biggest grassroots campaign in Arizona history" and raised over $2 million in support for Proposition 301" (White 10). The ballot was approved by voters by a margin of 53 percent to 47 percent and higher education institutions in the state will receive $45 million a year from sales tax for the next 20 years. Besides making these investments, Arizona's philanthropic community began to strategically think about the future of the economy and the promises of innovative industries such as biotechnology. In 2001, the Flinn Foundation began to focus on the bioscience economy in Arizona. The foundation issued a comprehensive study of the assets and opportunities in bioscience which in turn helped shape the investment priorities for the funding that resulted from Proposition 301.

The year 2002 was very eventful for Arizona's economy. A leading scientist who worked on the Human Genome project in Washington, D.C. decided to establish a non-profit research institute in Phoenix, Arizona. The so-called Translational Genomics Research Institute (TGen) works primarily on studies at the intersection of genomics research and the translation of findings into human health applications. TGen was founded by Jeffrey Trent, who grew up in Arizona and who was interested in returning. Once Arizona's political and economic leaders heard about his interest to move the institute to Phoenix, an immense community organizing and fundraising effort emerged. Within a very short time, a task force raised more than $80 million dollar (including funding from philanthropic groups such as the Flinn Foundation, Salt River Pima-Maricopa Indian Community, local and state governments among others). The City of Phoenix partnered with TGen on the development of a site in downtown Phoenix and TGen moved into its new $46 million building in 2004. The institute forms the cornerstone of the Phoenix Biomedical Center, a major downtown redevelopment effort which also includes the newly founded University of Arizona College of Medicine and Arizona State University's downtown campus.

In addition to these major investments, a leadership change at the state's largest university introduced a new kind of thinking about the role of universities in economic development. In 2002, Arizona State University hired a professor from Columbia University to become its 16th president. In the short time Michael Crowe has been president, he has introduced a new kind of thinking about the role of ASU in the local and global economy. His goal is to transform ASU into "one of the nation's leading public metropolitan research universities" (Arizona State University). Crow's philosophy is based on the model of the "New American University." He argues that universities like Arizona State cannot and should not try to imitate Ivy League universities like Harvard or Columbia. Instead, Arizona State should adopt new types of principles of social and economic engagement as well as access and research excellence. In his inaugural speech, Crow outlined several so-called "design imperatives" for a "New American University." The design imperatives illustrate how Arizona State is working to become more relevant in the region (community engagement, place making, economic development, and extended campuses in downtown and various suburban areas of the Phoenix region). They also emphasize a different thinking about knowledge creation and emphasize use-inspired research. In addition, the vision embraces interdisciplinary, multidisciplinary and transdisciplinary research and teaching. As a result of this imperative, ASU has built a new bioscience research institute (which collaborates extensively with TGen) that combines different fields and approaches to the subject. The university has also reorganized various departments and colleges. In addition, Arziona State University aims to combine research excellence with access to its university. Rather than limiting the number of students, ASU sees itself as an open university that is ready to take on different kinds of students. Such an approach is often seen at odds with an emphasis on research excellence and it will be interesting to see how ASU is working to overcome the challenges.

In 2006, the Arizona legislature created the 21st Century Competitiveness Fund and made $35 million available for science and technology projects. The funding is aimed towards creating a world-class science and engineering infrastructure in the state. The funds are administered by the Science Foundation Arizona (SFAz), which was modeled after the Science Foundation Ireland. Like other states, SFAz offers funding to leverage federal funds, to support graduate research fellowships, to seed investments in commercialization technology, and strategic partnerships between research institutions and industry.

Arizona represents a very interesting case. The state has adopted a leapfrog strategy and is trying to create a bioscience economy. Policymakers are utilizing research universities and research institutes as the building blocks. The state may, however, face stiff competition and some hurdles to achieving its goals. Virtually every state and major city in the United States is pursuing a bioscience or life science economic development strategy (cf. Battelle). What makes their tasks difficult is that only a select few metropolitan areas have been able to increase their share of commercial activity in biotech primarily because of the above average availability of venture capital, new firm formation, and the opportunities startup companies have to partner with larger pharmaceutical firms (cf. Cortright and Mayer). Arizona and specifically Phoenix may have a hard time building a biotechnology industry.

Conclusion

As the cases illustrate, aggressive efforts to improve a state's economic standing through investments in research and development are prevalent. Small and large states are adopting these kinds of public policies and policymakers hope to aid in the transformation of their economies. Even though these investments are small (mostly less than 3 percent) when compared to other sources such as private funding or federal funding in research and development, they signal a different approach to economic development. In particular, states are increasingly interested in being proactive agents in the transformation of their economies. Most often programs are used to provide incentives to faculty at local universities to leverage federal funding and to engage industry with university research efforts or vice versa. Very common is a focus on science and technology, particularly in biotechnology, high-technology, or nanotechnology. The focus on the so-called "nano, bio, info" sciences may, however, distract states from their historical economic and research strengths. It may on the other hand—if they are designed strategically—guide investments aimed at upgrading and transforming of traditional industries (cf. Lester). Lester argues that universities may play differentiated roles depending on the type of industries located in a region and state R&D funds should be sensitive to the local context and types of industries that could benefit from these investments.

State efforts are most often guided by a naïve believe that the benefits stay local and contribute to job creation and firm formation in-state. However, research has shown that innovative and entrepreneurial firms leverage programs across states and often piece together funding from different sources at the state and federal level (cf. Feldman and Kelley). Contrary to this corporate practice, most state efforts do not reach across boundaries even though a state's metropolitan economy may be geographically located in a multi-state metropolitan region such as in the case of Kansas City or Portland, Oregon. Political boundaries and allegiances prevent policymakers from making investments available to firms located outside of their jurisdiction. An exception might be Oregon's program to support nanotechnology research (ONAMI), which includes a research lab located in Washington State. Massachusetts's Research Center Funds also allow investments in collaborations between universities across state borders. States may want to rethink this practice especially given the emergence of global production and R&D networks.

Most state R&D investment programs emphasize in some way or another direct payoffs in form of commercialization and technology transfer. SFAz's Small Business Catalytic (SBC) program, for example, focuses on "seed" investments in research that has "high-impact commercial outcomes. SFAz awards are intended to leverage the potential of Arizona's researchers to secure much larger amounts of funding for technology commercialization. Companies created as a result of SFAz funding must remain in Arizona for a minimum of five years. The purpose of this investment is to create a catalyst for technology development, company formation and high-tech job creation in Arizona" (Science Foundation Arizona). The focus is on product and process development, rather than basic or even applied research. At American universities, applied research has a long history (cf. Rosenberg and Nelson). Before World War II, universities were linked in very practical ways to industry (primarily agriculture and forestry). These links still exist but they are smaller. In addition, during and after the war, universities began to build up their basic research expertise and the federal government recognized their capacity to conduct basic research, primarily in defense and health/life sciences. A parallel trend, however, has been the erosion of industrial R&D efforts and the decline of corporate research. Universities are often looked at to step into a void left by industry. Rosenberg and Nelson caution policymakers and argue that universities can contribute to advances in industry—as they have done for example in the engineering sciences—but that the division of labor between research and its commercial application has to be respected. In their view, universities can significantly contribute to research advances, but they are ill suited to create new products or processes.

In sum, the efforts states are undertaking to invest in research and development reinforce the trend towards the entrepreneurial university and are examples of the so-called triple helix of university-industry-government relationships (cf. Etzkowitz and Leydesdorff). Universities are becoming more pro-active in contributing to regional economic development and innovation. As in the case of Arizona State University, higher education institutions have become much more explicit about their contributions and are even redefining their missions. Arizona's "New American University" model illustrates this transformation well. State investments aid them in this transformation. Policymakers, however, may need to revise their naïve believes in the linear innovation model and adopt approaches that take into account that knowledge and innovation are created in a more dynamic and interactive fashion.

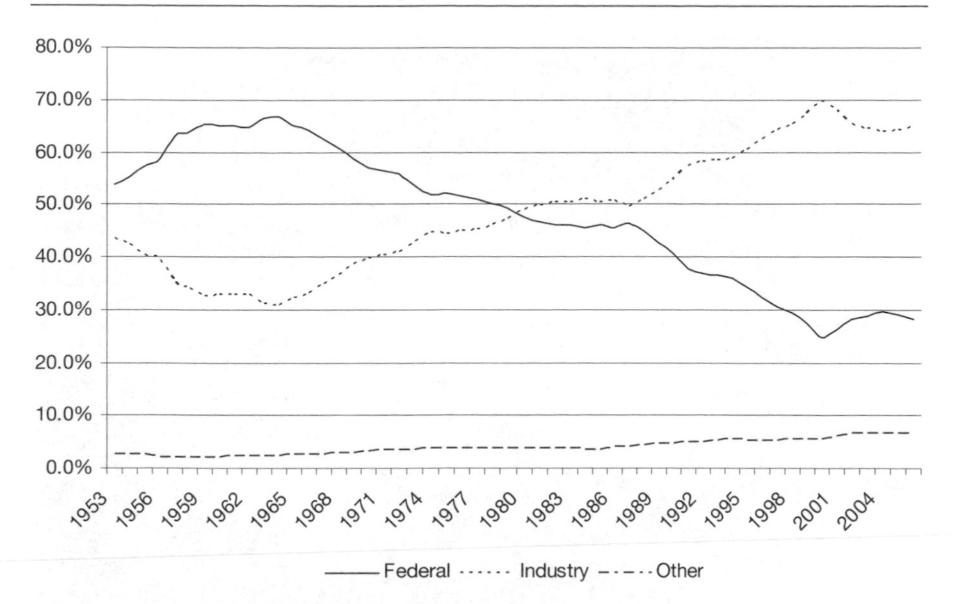

Figure 1: U.S. R&D expenditures, by source of funds: 1953-2006
Source: National Science Foundation, Division of Science Resources Statistics, National Patterns of R&D Resources (annual series)
Note: Data for 2005 and 2006 are projections

Figure 2: The Stowers Institute in Kansas City, MO
Source: Stowers Institute for Medical Research

Works Cited

Animal Health Corridor. "Kc Animal Health Corridor." 2007. Kansas City Area Development Council. 30 August 2007 <http://www.kcanimalhealth.com/>.

Arizona State University. "About Michael Crow." 2007. 20 September 2007 <http://www.asu.edu/president/meetthepresident/>.

Battelle. *Growing the Nation's Bioscience Sector: A Regional Perspective*. 2007.

Chesbrough, Henry. *Open Innovation: The New Imperative for Creating and Profiting from Technology*. Boston, MA: Harvard Business School Press, 2003.

Clarke, Susan E. and Gary L. Gaile. *The Work of Cities*. Globalization and Community. Ed. Dennis R. Judd. Vol. 1. 2 vols. Minneapolis, MN: University of Minnesota Press, 1998.

Cooke, Philip and Kevin Morgan. *The Associational Economy: Firms, Regions, and Innovation.* Oxford: Oxford University Press, 1998.

Cortright, Joseph and Heike Mayer. *Signs of Life: The Growth of Biotechnology Centers in the U.S.* Washington, D.C.: The Brookings Institution, 2002.

Davey, Monica. "Stem Cell Amendment Changes Little in Missouri." *The New York Times* 10 August 2007.

Dodds, Gordon B. and Craig E. Wollner. *The Silicon Forest: High Tech in the Portland Area.* Portland, OR: The Oregon Historical Society, 1990.

Etzkowitz, Henry and Loet Leydesdorff. "The Triple Helix of University-Industry-Government Relations: A Laboratory for Knowledge Based Economic Development." *EASST Review* 14.1 (1995): 11-19.

Feldman, Maryann P. and Maryellen Kelley. "How States Augment the Capabilities of Technology-Pioneering Firms." *Growth and Change* 33.2 (2002): 173-195.

Freeman, Chris and Luc Soete. *The Economics of Industrial Innovation.* Cambridge, MA: The MIT Press, 1999.

Gertzen, Jason. "Kansas as a Lab Finalist Reaches for the Top Tier." *Kansas City Star* 16 July 2007.

Gibbons, Jim. "2007 State of the State Address." Carson City, 2007. 27 August 2007 <http://gov.state.nv.us/PressReleases/2007/word/2007-01-22-StateoftheStateAddress.doc>.

Heartland Bio Agro Consortium. "Nbaf in Kansas." 3 September 2007 <http://www.nbafinkansas.org/task_force/>.

Kaiser, Robert. "Multi-Level Science Policy and Regional Innovation: The Case of the Munich Cluster for Pharmaceutical Biotechnology." *European Planning Studies* 11.7 (2003): 841-857.

Kovsky, Eddie. "Otter Recommendations Disappoint Tech Council." *Idaho Business Review* 22 January 2007.

Lester, Richard K. "Universities, Innovation, and the Competitiveness of Local Economies: Summary Report from the Local Innovation Project – Phase I." 2005. MIT Industrial Performance Center. 28 December 2006 <http://web.mit.edu/ipc/publications/pdf/05-010.pdf>.

Maryland Industrial Partnerships. "Mips: Success Stories." 2007. University of Maryland. 23 August 2007 <http://www.mips.umd.edu/success_stories.html>.

Mayer, Heike. "Competition for High-Tech Jobs in Second Tier Regions: The Case of Portland, Oregon." *Building the Local Economy: Cases in Economic Development.* Eds. Douglas Watson and John Morris. Athens, GA: Carl Vinson Institute of Government/University of Georgia, 2008.

---. "Completing the Puzzle: Creating a High Tech and Life Science Economy in Kansas City." 2006. The Brookings Institution <http://www.brookings.edu/metro/pubs/20061101_kansasmayer.pdf>.

---. *A Review of State R&D Investment Funds: Ten Case Studies.* Alexandria, VA.

---. "Taking Root in the Silicon Forest: The Role of High Technology Firms as Surrogate Universities in Portland, Oregon." *Journal of the American Planning Association* 71.3 (2005): 318-333.

Morrison Institute for Public Policy. “Arizonans’ Attitudes toward Science, Technology, and Their Effects on the Economy.” Phoenix, 2006. Arizona State University. 9 September 2007 <http://www.asu.edu/copp/morrison/SciSurFNL.pdf>.

National Governors Association, and Pew Center on the States. "Investing in Innovation.” Washington, D.C., 2007. 27 August 2007 <http://www.nga.org/Files/pdf/0707INNOVATIONINVEST.PDF>.

New York Times. “Stem Cell Proposal Splits Missouri G.O.P.” *The New York Time.* 12 March 2006.

Oinas, Paivi. “Finland: A Success Story?” *European Planning Studies* 13.8 (2005): 1227-1244.

Oregon Council for Knowledge and Economic Development. *Renewing Oregon’s Economy: Growing Jobs and Industries through Innovation. A Report from the Oregon Council for Knowledge and Economic Development.* Portland, 2002.

Oregon InC. “2007 Innovation Plan.” 2007. Oregon Innovation Council. 30 August 2007 <http://www.oregoninc.org/InnoPlan.pdf>.

Ostendorf, Berndt. “A Nation with the Soul of a Church? The Strange Career of Religion in America: A View from Europe.” *Rivista di Studi Nord Americani* 15/16.Special European Issue (2005): 169-196.

Pew Center on the States, and National Governors Association. “Investing in Innovation.” Washington, D.C., 2007. National Governors Association. 23 August 2007 <http://www.nga.org/Files/pdf/0707INNOVATIONINVEST.PDF>.

Plosila, Walter. “State Science-and Technology-Based Economic Development Policy: History, Trends and Developments, and Future Directions.” *Economic Development Quarterly* 18.2 (2004): 113-126.

Rosenberg, Nathan and Richard Nelson. “American Universities and Technical Advance in Industry.” *Research Policy* 23 (1994): 323-348.

Science Foundation Arizona. “Grants & Investments.” 2007. 21 September 2007 <http://www.sfaz.org/investments.html>.

Sternberg, Rolf and Christine Tamasy. “Munich as Germany’s No. 1 High Technology Region: Empirical Evidence, Theoretical Explanations and the Role of Small Firm/Large Firm Relationships.” *Regional Studies* 33.4 (1999): 367.

Waits, Mary Jo. “The Added Value of the Industry Cluster Approach to Economic Analysis, Strategy Development, and Service Delivery.” *Economic Development Quarterly* 14.1 (2000): 35-50.

White, Bradford. *A Case Study of Proposition 301 and Performance-Based Pay in Arizona*: Consortium for Policy Research in Education, 2002.

From Black Box to White Box: The Changing Life Worlds of Communication Technologies

David E. Nye

The following provides a meta-commentary on the changing perception and use of communication technologies between c. 1850 and the present, and is intended to provide a framework for further research. My basic methodological assumption, widely shared by scholars in the history of technology, is that machines are socially constructed, both in their invention and their use (cf. Bijker). When devices such as the telephone, the phonograph, the radio, or television emerge, they do not automatically have a particular impact. Rather, each machine is incorporated—or, in some cases, rejected, like AT&T's picture phone (cf. Lepartito)—in a process that may take years (Nye 2006, 33-48).

Rejecting technological determinism, however, does not mean that machines are infinitely malleable or that they can become, or mean, whatever consumers would like them to. Technologies engage physical constraints that cannot always be overcome. For example, as early as the 1890s inventors realized that batteries were heavy and took too long to charge, and for a century both of these problems proved intractable (Israel 410-421). During this time, electric automobiles could not compete very well against gasoline engine cars, which were lighter, could be refueled in a matter of minutes, and had a greater range. Likewise, American consumers of the 1930s wanted inexpensive family airplanes, but it was not possible then or now to lower costs enough to reach this market. Products are shaped in part by the physical limits and costs of materials, as well as by the level of engineering knowledge at a particular historical moment (Petroski 3-27). While the rest of this paper will not dwell on these matters, it is written with a conviction that only within technical limits can inventors, manufacturers, and consumers influence the forms of their machines.

Furthermore, if communication devices are not deterministic, neither are they so flexible as to become fluid expressions of contemporary human desires (cf. Friedel). The forms of inventions often stabilize in a form that is not optimal, but which remains fixed both from habit and due to the high cost of change (David 332-334; cf. Puffert). Once embedded in social life, machines embody earlier consumer choices and often are slow to change. After a technology is established, manufacturers want stability in product design so they can focus on manufacturing and marketing (cf. Utterback). Consumers also need time to adjust to new technologies such as the phonograph and telephone, and are often content with incremental improvements in performance, such as better sound quality, rather than seeking radical design changes. Technologies of communication do not change at roughly the same rate year after year. Rather, there are periods of stability, with a succession of small improvements (the telephone, c. 1900 to 1980) and periods of rapid transformation, when the design, function, and use of devices all change radically (the telephone after c. 1980).

As was the case with the telephone, a major shift has occurred during the last quarter century in the ordinary users' attitudes toward and experience of all communication technologies. As late as the 1980s, an older view was not merely common, but almost universal. Until roughly that time the major forms of communication, notably the telephone, phonograph, radio, television, and computer shared a set of physical characteristics. I will refer to this system of machines and the attitudes toward them as "Black Box" technologies. In the last quarter century, however, all of these objects have undergone a radical miniaturization and become portable. In the process, these communication forms have assumed a new relationship to those who use them. I will refer to these new miniaturized objects as "White Box" technologies.

While the most obvious difference between these two sets of objects is their size, it is important not to rush past this aspect of the contrast. The earliest of these technologies, the telephone, from 1876 until the 1980s was a heavy object. At first it was often fixed to the wall, or it sat permanently on a desk. For more than a century each telephone was tethered to a particular spot, and to use it required that one go to particular places. For the first two generations of its use, from 1876 until the middle of the 1920s, the telephone diffused only slowly into society, and the average home did not have one. People had to search for a telephone box or find someone who would permit them to use a private phone. The immobility of the heavy technological object was an aspect of its character, whether the phonograph, radio, television, or early computer. For decades after being introduced, each was large and heavy, and each was anchored to a power source or a communication line in the nearest wall. It was inconceivable that one could misplace any of these objects, and none was portable. They were fixed parts of domestic and office space. Phonographs, radios and televisions were massive pieces of furniture, often encased in wood. Their owners displayed them proudly and typically placed them in the most central and visible rooms of the house.

As historians of technology often noted, however, the owners of these machines seldom understood how they worked. The public treated them as "Black Boxes" whose innards were a mysterious tangle of tubes and wires that they learned not to touch. Manufacturers usually attached a warning label to the back of the cabinet of a phonograph, radio, television or computer, telling owners how dangerous it was to attempt any repairs themselves. When a curious child pried off the back and looked inside, there were few moving parts, and little to explain how they worked. In contrast, older, pre-electric technologies had gears, levers, wheels, and other moving parts. One could, with some diligence, understand the mechanical system that drove a music box or a clock. If it broke, there was often little danger if one took it apart, and many people handy with tools could make repairs.

In contrast to these mechanical devices, most people did not know much about how or why electrical technologies worked. At first, they seemed wonders that bordered on the supernatural but as they became familiar and gradually were naturalized as permanent parts of the everyday world, people stopped thinking about how they worked. This process of naturalization took a generation for each technology, starting with the phonograph and telephone, both of which seemed miraculous in the 1870s, but had passed into fairly common use a quarter century later. Children born in the twentieth century accepted them as unremarkable parts of everyday life, as objects that were always already there. The same occurred later with the radio, which seemed amazing in the 1920s, became a central form of popular culture in the 1930s and 1940s, and then faded into the ordinary after 1950. At that point, excitement shifted to another Black Box technology, the television. Currently, Black Box technologies are being eclipsed, as they are combined and absorbed by digital White Box machines.

What were the other major characteristics of these Black Box technologies? With the exception of the early hand-cranked phonograph, they were electrical. They required that the consumer either had a powerful battery or was linked to the utility network. It is easy to pass over this point, as it may seem almost trivial. But to the American consumer between 1890 and 1930, the arrival of electricity in the home and office was experienced as a major revolution (Nye 1990). Electricity illuminated the spaces of daily life far more brilliantly than gas jets or kerosene lanterns, and the centrality of the hearth diminished. Electricity homogenized space, making it possible to bring power, heat, light, and communications to any location. At the same time, these early networks of wires accustomed the public to establishing permanent communication links to the larger society. By 1930, Americans had not only accepted and largely naturalized the particular technologies of the telephone, phonograph, and radio, but they had also absorbed the idea of being enmeshed in a vast system of communications that was inseparable from the electrical infrastructure. In this way, the Black Box attitude toward the individual technologies was generalized to the network as a whole. For if the consumer had little notion of how any of the individual machines functioned, they had even less comprehension of the interlocking grid that linked their local utility into complex regional power systems. From this perspective, adding the television to everyday life (for the consumer it was essentially radio merged with film) required less mental adjustment than earlier had been necessary to accept the emergence of the electricity-based communications system. Consumers focused on the content and the entertainment, but made little attempt to understand how radio or television worked.

This Black Box system had other shared characteristics. Each of these objects had a single function. At times, a manufacturer would package two of them in one cabinet, notably the phonograph and the radio, but they remained two distinct machines, sharing a loudspeaker system to be sure, but the synergy ended there. The original Edison phonograph of 1877 could both record and play back, but this feature was suppressed over time, and the phonograph/radio combinations generally did not allow consumers to record from one to the other. Because these technologies were not inter-operable, the Black Box system permitted a vast diffusion of popular music, but still preserved copyright during most of its dominance. The technologies were each quite separate, and the boundaries between them really did not begin to break down until the 1970s, with cassette radio phonographs. Until then, if one wanted a hit song, it had to be purchased as a record or the radio station had to pay a royalty to broadcast it.

These single-function machines were analogue rather than digital devices. They had mechanical controls. One turned a knob or pressed a button and heard a click that signaled the machine had come on. For more volume, one turned another knob, and there was a direct correlation between that physical action and the result. Inside their cases, individual tubes and wires could be replaced. Furthermore, these technologies of sound and visual reproduction did not change all that rapidly. An AT&T telephone from 1925 often remained in service for twenty years. A radio from the 1930s could still be used in the 1950s. A phonograph record from 1925 could still be played in 1960. In contrast, in digital technologies software size rapidly outgrew memory, forcing constant upgrades. A PC from 1990 was outmoded by 1994, and completely unusable in much less than a decade. Black Box technologies were relatively stable. Such objects were expected to work for a generation. Although improvements were made to later models, there was less urgency about buying replacements.

Black Box communications were used as much collectively as individually. When radio first emerged in the 1920s whole families would gather around the set and marvel at hearing songs and clever conversation from distant locations. The sound of radio soon permeated homes and many work places, and while this is often called "mass communication," in fact local stations held their own against the national networks. Only local stations could cater to the advertising needs of local businessmen, and only local radio could forge a strong link to listeners through local call-in programs, competitions, and the like (Cohen). Likewise, early telephones were generally party-line devices, shared by as many as three or four families (Fischer). Whenever they rang, several people might pick up and listen in. Furthermore, operators literally made the connections. In a small town, they knew all the customers, and in emergencies people would call the telephone office to locate a person who was not home. The human switchboard operators at the nerve center of the system generally heard any important gossip. Between 1880 and 1950, such party lines and operators increased the sense of neighborliness and community. When AT&T made area codes and direct dialing available in the early 1950s, however, the telephone rapidly lost this community function. At the same time, the party line was being replaced by private phones, which were necessary in order to know what family made which direct-dial, long distance calls.

In the early 1950s, however, the newer Black Box technology of television still drew neighbors, parents and children together. In the first years, televisions were still too expensive for many families and no household had more than one. Collective viewing was particularly common among children, who liked to see such programs as "The Mickey Mouse Club" or cartoon shows together. Teenagers tended to monopolize the television in the late afternoon, when they saw "American Bandstand" and similar programs. They usually did more than watch. From television, they learned the latest dance steps as well the lyrics to new songs. Many adults also watched television together. Men particularly enjoyed baseball and football games, which in fact often induced them to buy a television in the first place. Networks catered to housewives with soap operas and in some cases the women formed groups around particular shows. When characters got married, some fans dressed up and celebrated as though they were going to a real wedding. Television also brought people together in another way, for most saw the same programs, whether together or separately. There were few channels and the relatively few hours of programming were concentrated in the late afternoons and evenings. There were only three national networks, but in many parts of the country just one or two of them came in. I grew up in central Pennsylvania, hardly the most remote area of the United States, yet in 1955 we received just two stations, only one of them clearly. Given such a limited choice, neighbors frequently saw the same programs, and the following day it was common to share opinions about what they had seen. Because people shared party line telephones, clustered around the radio, or grouped themselves around televisions, the interface between consumers and these devices was collective. It also was often a loud interface, as a group watching or listening together talked and laughed and turned up the volume.

The implicit contrast to all of these observations by now is clear. The White Box technologies are not fixed, but portable. They do not weigh 20 kilos or more, but can be carried in a shirt pocket. They are scarcely mechanical in any way, but electronic. They are digital, not analogue devices. They are multifunctional, not single functional. The newest devices on the market combine the telephone, television, radio, phonograph, and computer. They are designed not for group use and the creation of shared immediate experience, but for individual use and for deferred sharing with a virtual network of distant others. They require electricity to work, but they are not permanently linked to the network with wires, for the whole system is—and consumers feel it must be—portable. White Box consumers are not publicly loud, but wear earphones that shut them off from others.

The White Box technologies intensify some trends that began with the Black Box devices. If the Black Box was hard to understand, these new machines are entirely inscrutable, and cannot be understood as a system of clearly separated parts. It is usually difficult to pry off the back or the side, but if one manages to take a computer or an Ipod apart, one finds tiny solid state components that a professional may be able to replace but which themselves can almost never be repaired. There are no longer any moving parts. The cost of repairs is high, assuming one can even find a technician who will make them. The obsolescence of these products comes far more rapidly than with the Black Box technologies. Adult consumers trade in mobile phones within a year or two, but people under 20 change do so more than once a year. The mechanical typewriter could withstand thirty years of pounding service, but in businesses computers are quickly judged obsolete, even though they still work, and replaced after two or three years. The Black Box technologies were all hardware, but the digital White Box technologies can become obsolete either in terms of software or hardware. They add an expensive intermediary layer of coding that also can be corrupted, sold, pirated, or upgraded, eroding the sense of ownership and of control.

White Box technologies are replaced more often in part because new functions are constantly being added. They therefore demand more consumer expertise as well. An old-fashioned radio had two knobs, one to turn it on and control the volume, and one to search for stations. An old-fashioned telephone had a speaker, receiver, and a dial, nothing more. In contrast, even a quite ordinary cell phone has a much larger range of functions. Many of the Black Box machines merge into a single device that contains at the very least a camera, a calculator, a calendar, an alarm clock, a telephone, SMS messaging, and often a radio and music player. More advanced models literally contain all of the Black Box functions, including television, web browser, and some computing functions.

People once used Black Boxes to effect a major change in social life, as they moved social interaction and entertainment out of the public realm into private, domestic space (Nasaw). In contrast, White Boxes take communications out of any fixed space, and make them part of a new individualistic nomadism. The people one sees in the streets and cafes with their mobile phones and Ipods are intensely engaged with sounds and conversations that distract them from their surroundings. They can be irritatingly oblivious to others, as they turn inward to their own private soundscape. They ignore aural and visual surroundings. On airplanes and trains they seldom look out the window or see their fellow passengers, because they are busy sending SMS messages, playing miniaturized computer games or viewing films or prerecorded programs, These figures journey through the postmodern landscape without paying attention to it, accepting its fragmentation and iconographic discontinuity as the appropriate background for their channel surfing, interrupted by frequent text and voice messages.

The figure in the romantic landscape of nature gave surroundings undivided attention, seeking to be absorbed. Those who took the train to visit the Grand Canyon or Niagara Falls often stayed for a week. That attitude and sensibility began to erode as the Black Box technologies emerged after 1875. The figure in the modern landscape celebrated the technological sublime (Nye 1992), in which engineers demonstrated their ability to conquer space and time (cf. Harvey). The telegraph, when first invented, was considered a wonder. Americans likewise embraced the railroad as sublime, and celebrated each new communication technology because it accelerated and intensified their experiences. This compression of space and time pushed the individual into greater contact and identification with a human-built physical world. The narration of experience assumed a double quality, as the tension increased between the romantic conception of the natural world and the technological transformation of experience through Black Box technologies. So long as these remained single function machines that were fixed in space and shared by multiple users, individuals mediated the tension between the world and the new forms of communication.

White Box technologies are not about mediation, but intermediation. The reference point is not a natural world being conquered and surpassed, but rather intertextual referentiality. The natural world ceases to be the point of departure, replaced by a world of simulacra, with no ultimate reference point. "The simulacrum is never that which conceals the truth – it is the truth which conceals that there is none" (Baudrillard 162). An endless chain of reference without beginning or end is embedded in everyday experience, for example in the sampling in popular music or the visual quotations on YouTube. At the same time, the ideal individual is not absorbed in one thing, but a multi-tasker, a juggler, capable of simultaneously doing several things, one of which, however, is not likely to be contemplating the landscape or communicating with the people nearby. This new figure is not really in the landscape at all. This person with the mobile phone, ideally at this moment an Apple Iphone, is passing to and fro but not really part of any social space. The local and immediate is literally screened out. Instead, this new figure in the landscape of the White Box world has embraced cyberspace. The consensual hallucination described by William Gibson in *Neuromancer* seems to approach, not only in the form of cyberspace but also in an overlay of electronic icons that guide and interpret navigation through the "real" world.

There have been critics of all these developments. The shift away from the Romantic view of nature deeply concerned the Transcendentalists, John Muir, and more recently the environmental movement. The tension between *The Machine* and *The Garden* also provided the theme of Leo Marx's classic work in American Studies. In focusing on the railroad and nineteenth century literature, his work concerns the early form of industrialization that came with the steam engine. The Black Box technologies emerged later, and are part of a different cultural configuration, that was a transition from the legible, mechanical devices of the industrial revolution in c. 1850 to the opaque White Boxes of 2000.

Where industrialization required mechanical forces such as falling water, steam, or burning gasoline, the Black Box technologies required electricity, and White Box devices require portable electricity and wireless communication. This change in the sources of power effectively measures how devices became divorced from any direct connection to the world around them. The power of the steam engine came from a visible and noisy source. The energy for the telephone or television flowed silently through a cable. But mobile devices are only grudgingly plugged in to recharge, for ever shorter times as batteries improve. In 2005 Lawrence Ferlinghetti suggested that, "What's called the dominant culture will fade away as soon as the electricity goes off." He identified the electrified world with the evasion of immediate reality, recalling that, "In the 60s there was a famous slogan, 'Be Here Now,' which in fact was a best-selling book by Ram Dass. Today, with the telephones, the fax, the Internet, the whole schmear– the slogan you have today is 'Be Somewhere Else Now.'" By extension, one might say that in the displacements of the White Box world, there is no "here." Once daily lives are intertwined with the computer and mobile telephone people often feel it is impossible to get anything done without them. The immediate physical surroundings are insufficient.

Ferlinghetti's critique has roots in European literature and philosophy, as well as in American Transcendentalism. Henry David Thoreau lived before most of the electrical inventions, but he cast doubt on whether new technologies are always an advantage and questioned whether the telegraph was really an improvement (Thoreau 42). Likewise, in Max Frisch's novel *Homo Faber*, an electrical engineer in Latin America installs electric power stations, which help to "eliminate the world as resistance." Electric light eliminates the night, air conditioning eliminates climate, and electric devices replace physical labor such as pumping water or grinding wheat into flour. The collective result of such technologies is that people lose their direct experience of the world.

Similarly, beginning in the 1930s Martin Heidegger and other philosophers argued that in highly technological cultures people cease to feel connected to the natural world. Although they are part of nature, they use modern technologies to arrange their lives so that they scarcely experience it. One of Frisch's characters complains of "The technologist's mania for putting the Creation to a use, because he can't tolerate it as a partner, can't do anything with it; technology as the knack of eliminating the world as resistance, for example, of diluting it by speed, so that we don't have to experience it" (Frisch 178). As White Box technologies mediate experience, they erase immediate knowledge of and contact with the physical world.

This Heideggerian critique may be extended further. As a web of White Box intermediation (re)defines "normality," and as this web becomes more complex with each generation, humankind will construct an entirely new "normality." Postmodernists such as David Harvey emphasized how the acceleration of transportation and communication compressed space, sped up time, disconnected voice from presence, subverted social boundaries, intensified the circulation of information, and created a blizzard of representations. Only a placeless, postmodern self can accommodate this nomadic, multitasking infrastructure. Simultaneously, as Steven Connor has suggested, "the reconfiguration of the relations between the senses, especially of hearing, seeing and touch, promised by new communicative and representational technologies may allow for a transformation of the relations between feeling, thinking, and understanding . . . . The sheer overload of sensory stimulus required to absorb by eye and ear results in a switching or referral of senses. These contemporary synesthesias make it appropriate for us to think in terms of visual cacophony and white noise . . . . we must also expect a redistribution of the values previously sedimented in the senses of hearing, vision, taste, touch, and smell" (Connor 162). The experience of being alive, to the considerable extent that it is defined through the senses, has already changed fundamentally. The new prescription for human beings seems to be that, like their White Boxes, they too must be multi-functional devices that seldom need be tied to any location.

The consequences of using White Box technologies for community formation and social solidarity are unclear. Some argue that the sense of social contract is dying, and that contemporary people are isolated, "bowling alone" (cf. Putnam). Indeed, the fear that modern communications would paradoxically lead to personal isolation despite being immersed in a sea of messages had already emerged before World War I (cf. Forster). More hopefully, others claim that the White Box technologies make possible "smart mobs" (cf. Rheingold) who share information, link up on FaceBook, mobilize quickly, and raise unheard of sums for the political candidates of their choice. It is too soon to know how the White Box technologies will be used, but certainly the Black Box machines that once seemed revolutionary have become antiques.

Works Cited

Bijker, Wiebe E., Thomas P. Hughes, and Trevor Pinch (eds.). *The Social Construction of Technological Systems: New Directions in the Sociology and History of Technology.* Cambridge, MA: MIT Press, 1987.

Cohen, Lizabeth. *Making a New Deal: Industrial Workers in Chicago, 1919-1939.* Cambridge, MA: University Press, 1990.

Connor, Steven. "Feel the Noise: Excess, Affect and the Acoustic." *Emotion in Postmodernism.* Ed. Gerhard Hoffmann and Alfred Hornung. Heidelberg: Universitätsverlag Winter, 1997.

David, Paul. "Clio and the Economics of QWERTY." *American Economic Review* 75 (1985): 332-337.

Douglas, Susan. *Inventing American Broadcasting, 1899-1922.* Baltimore, MD: Johns Hopkins University Press, 1987.

Fischer, Claude. *America Calling: A Social History of the Telephone to 1940.* Berkeley, CA: University of California Press, 1994.

Forster, E. M. *The Machine Stops and Other Stories.* Ed. Rod Mengam. London: Andre Deutsche, 1997.

Friedel, Robert. *The Culture of Improvement.* Cambridge, MA: MIT Press, 2007.

Frisch, Max. *Home Faber.* New York: Harcourt Brace and Company, 1987.

Gibson, William. *Neuromancer.* Glasgow: HarperCollins, 1984.

Harvey, David. *The Condition of Postmodernity.* Oxford: Basil Blackwell, 1989.

Heidegger, Martin. *The Question Concerning Technology and Other Essays.* New York: Harper and Row, 1977.

Israel, Paul. *Edison. A Life of Invention.* New York: Wiley, 1998.

Lepartito, Kenneth. "Picturephone and the Information Age: The Social Meaning of Lyotard Failure." *Technology and Culture* 44.1 (2003): 50-81.

Marx, Leo. *The Machine in the Garden.* Oxford: University Press, 1965.

Nasaw, David. *Going Out.* Harvard, MA: University Press, 1999.

Nye, David E. *Technology Matters: Questions to Live With.* Cambridge, MA: MIT Press, 2006.

---. *American Technological Sublime.* Cambridge, MA: MIT Press, 1994.

---. *Electrifying America: Social Meanings of a New Technology.* Cambirdge, MA: MIT Press, 1990

Petroski, Henry. *Small Things Considered: Why There is No Perfect Design.* New York: Vintage, 2003.

Poster, Mark (ed.). *Jean Baudrillard. Selected Writings.* Stanford, CA: University Press, 1988: 166-184.

Puffert, Douglas J. "Path Development in Spatial Networks: The Standardization of Railway Track Gauge." *Explorations in Economic History* 39 (2002): 282-314.

Putnam, Robert D. *Bowling Alone*. New York: Simon and Schuster, 2000.
"Questions for Lawrence Ferlinghetti." *New York Times Magazine* 11 (2005): 19.
Rheingold, Howard. *Smart Mobs: The Next Social Revolution*. New York: Basic Books, 2003.
Thoreau, Henry David. *Walden*. New York: Holt, Rinehart and Winston, 1948.
Utterback, James M. *Mastering the Dynamics of Innovation*. Boston, MA: Harvard Business School Press, 1994.

A Signal Success: An Illuminating History of One Woman, One Invention

Denise E. Pilato

Martha J. Coston (1828-1904), a successful inventor during the 19th century, patented a pyrotechnic signal flare and international code system. She engaged in the business of invention and in the manufacturing and marketing of her signal flare the establishment of a long and prosperous business, the Coston Supply Company. The first *successful* utilization of the Coston Signal occurred during the United States (U.S.) Civil War (1861-1865) by the U.S. Navy. They were later used extensively by the U.S. Army, the U.S. Weather Bureau, merchant vessels, and particularly by the U.S. Life Saving Service, which later became the U.S. Coast Guard. Over a 15-year period, beginning in 1859, Coston also *successfully* marketed her signal and code system in England, France, Holland, Austria, Denmark, Italy, and Sweden. The Coston Signal became a ubiquitous form of communication with special effectiveness in darkness and bad weather. Coston Signals proved invaluable as a lifeline to many, including the inventor.

In addition to the *successful* invention of the signal flare, Coston invented an unusual identity for herself. During an era when the prototype image of an inventor was implicitly male, she emerged as a professional female inventor and businesswoman. The idea of a woman inventor presented something of a sort of paradox in the 19th century, but there was nothing self-contradictory in Coston's identity validated by her experiences as an inventor. She was committed to the success of the Coston Signal motivated by a "heartfelt desire to accomplish something for the good of humanity," and, of course, money (Coston 3).

In 1886, Coston published her autobiography *A Signal Success: The Work and Travels of Mrs. Martha J. Coston*, which provides reliable clues to documented sources verifying her *success*, as well as her obsession with it. She knew that although the signals were absolutely *successful*, "small minded" men begrudged a woman her success (Coston 272). What were the significant differences in the measure of success of the invention verses the inventor? Coston's experiences as a 19th century woman inventor represent a shifting paradigm not only in the gendered idea of inventor identity, but in the reality of female accomplishment despite gender discrimination, educational limitations, lack of business acumen, and restricted access to commercial networks. Coston approached her chosen profession with determination and persistence maintaining a clear objective of *success* at every major turn. Still, her name, her contributions, and her place in the history of technology remain clouded in obscurity.

I was first introduced to Coston, her remarkable story, and some of the underlying reasons why she and other important women inventors have been excluded in texts and dialogue relevant to the history of technology while working as an exhibit research team member at a major American history of technology museum in the early 1990s. My assignment was to research women inventors in an effort to find some who might be included in an exhibit focused on things made in America. During a three-month research period, I examined patent records found in the government census *Women Inventors to Whom Patents have Been Granted by the United States Government 1790 to July 1, 1888,* which includes over 5,000 patents awarded to women. After following up on inventors with potentially good stories, my recommendations included 12 women, whom I ranked based on various criteria such as artifact accessibility, longevity of impact of the invention, technical progress, and social interest. I was astounded to learn that ultimately not one woman inventor would make it into the exhibit. The reason why? It was deemed by the exhibit team leaders that none of the women inventors constituted a significant story. More than a decade later, that response still disturbs and motivates me. A simple driving question in my research has been, "what constitutes a significant woman inventor's story?"

Martha J. Coston's story offers a model of significance not only because she was a successful inventor, but also because her relationships with American and European military and government agencies document multifaceted experiences as an inventor. Her story also includes compelling human interest as a single, working mother in 19th century America. In the preface of her only published writing, she tells the reader that her purpose in writing was not because she was an author of any merit, but in part, because she thought that her story would "encourage those of my own sex who are stranded upon the world with little ones looking to them for bread, may feel, not despair but courage" (3). There are certainly other significant women inventor's stories. Some are significant because of high profile legal battles or bizarre strategies for negotiating gender discrimination, but the significance of Coston's experience is in her legacy of success. Both her product and identity as inventor were notable successes, technically, financially, and socially.[1]

[1] Margaret Knight is credited with anywhere from 22-89 inventions. She successfully fought a patent infringement lawsuit in 1870 over her invention of a machine that made the bottom of a paper bag square. Amanda Theodosia Jones, who invented a vacuum process for food preservation and other notable mechanical devices, was aided by an unseen spiritual mentor who

Concepts of success in the history of technology are inextricably connected to the notions of progress. How one measures success is often relative to trade-offs and consequences related to social and technical progress. For inventors, generally, success is constituted by profitability, technical accomplishments, professional recognition, personal growth, social values, or other mediating factors pertinent to the nature of their inventive work. By examining Martha J. Coston's experience in the U.S. Navy during the Civil War, her international business marketing strategies, her contributions to the U.S. Life Saving service, and the establishment of her long-standing business, a significant story of success is illuminated.

The Early Years

Coston was born in Baltimore, Maryland in 1828 and grew up in Philadelphia, Pennsylvania. In 1844, at 16 years of age, she married Benjamin F. Coston, a young and promising inventor. By 1848 she was widowed with four small children. When her husband died, she found herself ill prepared to take care of herself and her children. She admitted that she "knew not how to dig" and was "ashamed to beg" (Coston 37-38). She lamented that nature had not bestowed on her "a little of that brilliant genius so liberally given to my husband" who, at the time of his death, was credited as the inventor of the percussion cap, a type of rocket, percussion primers for cannon, the "Infernal Machine," which was an early submarine, and the Lanyard lock.[2] At the time of his death, he was working on several different invention ideas including experiments with gas lighting, which contributed to his illness and death. After his death, Coston had the idea that she might patent some of his unfinished inventions to support herself and her family.

Two years passed and the widow Coston was still steeped in numbing grief over the loss of her husband, which had been followed one month later by the death of her youngest son, Edward, and shortly thereafter by the death of her beloved mother. By 1850 on a "dreary November afternoon" while going through a trunk of her husband's things she found "numerous packets carefully sealed and labeled." They contained "unfinished inventions, inventions too costly to be utilized." She also found evidence of successful pyrotechnics experiments. When she came upon one large envelope containing ideas for signals to be used at sea and in the night, she recalled that she had discussed the idea of the flares with her husband during their courtship during which time he had made a few test signals at the

spoke to her and related technical details (Pilato 120-121, 129; Macdonald 54-55; Stanley 520-521).

2 Benjamin F. Coston's inventions were widely adopted by the U.S. Navy. He died of complications related to toxic poisoning associated with gas light experimentation in 1848 (Laurel Hill Cemetery, Philadelphia, PA. Martha J. Coston, file).

Washington-Navy Yard. She immediately sent a letter requesting the return of the signal samples. After many delay tactics, the samples were finally returned to her in a dilapidated state with the added note that the Navy never received "any recipes from Mr. Coston" on composition or manufacturing process. The Navy admitted that many attempts had been made in an effort to discover the missing composition and process, but without success, which accounted for the delayed return of the samples and their bad condition (37-40).

Her initial contact with the Washington Navy-Yard put into motion events that obsessively occupied the next decade of her life. During those first ten years, she faced difficult challenges, but not impossible obstacles. One technical problem that threatened the success of the signal was that the signal code system required three colors. She successfully developed a "vivid red" and "pure white" and wanted blue as the third color "in order to use the national colors" (44). However, due to her limited knowledge of chemistry and lack of experience in scientific methodology, the third color remained illusive. By 1858 she was desperate for a solution as the Navy testing of signals continued to be unsatisfactory and her goal was to sell the patent to the U.S. Navy. The solution came to her in an unexpected way after watching the New York celebration of the laying of the first transatlantic cable in 1858, which included spectacular fireworks. She reasoned that someone who could produce brilliant colors for fireworks could help her produce a bright blue. She began corresponding with several of the New York pyrotechnists under a man's name fearing that they would not give heed to a woman.

Mr. J. J. Detwiller responded and stated that he had made a blue color some years previous. Coston urged him to duplicate the blue, but if not, she would be interested in a strong green. Within ten days, she received a package containing a strong green color. In the end, her desire for the patriotic red, white, and blue could not be achieved with the same clarity and brilliance as green. Coston immediately entered negotiations to work with Lilliendahl-Detwiller Fireworks Factory in Greenville, North Carolina (44-46).[3]

Finally, in early 1859 she received notice of successful testing of the signal by the U.S. Navy under a specially appointed board of Naval Examiners by Secretary Toucey. A report published by the Navy Examiners in February, 1859, concluded that, "The Application of the "Coston night signals" to the navy day signal books gives a perfect code of night signals. They offer precision fullness, and plainness, at a less cost for fireworks than it is thought we now pay for confusion and uncertainty." In the opinion of the Board of Examiners, the signals were "decidedly superior" to any signals that the Navy had ever used ("Coston's Telegraphic Night Signals").

[3] Personal correspondence with Frederic C. Detwiller, great grandson of J. J. Detwiller (June 2004).

After nearly ten years of experimentation, negotiation, and persistence, Coston was awarded Patent No. 23,536 on April 5, 1859, for a pyrotechnic night signal and code system. She was listed as Administrix on this first patent, with her husband named as inventor even though he had been dead for over 10 years. Her second patent for "Improvement in Pyrotechnic Night Signals" filed in 1871, was under her own name as inventor. In part, this change of patent ownership and identity as inventor was influenced by her professional relationship with the U.S. Navy throughout the Civil War and her successful international marketing experiences, which she did entirely on her own (Pilato 84).[4]

U.S. Navy & The Civil War

Coston was in Europe marketing the signals in 1860 when "mutterings of war" in the United States reached her. She sailed for home and went directly to Washington, D.C. with the intention of bringing a bill before Congress for the sale of her patent. She was confident that her signals would "prove a valuable auxiliary" for the Navy in the event of war because the signal could be "seen at a distance of fifteen or twenty miles, and in the fiercest gales of wind and rain at a distance of several miles" (Coston 84).

She arrived in time to attend Abraham Lincoln's presidential inauguration on March 4, 1861, at the Willard's Hotel in Baltimore. On April 19, 1861, President Lincoln issued the Proclamation of Blockade of Southern ports, which reached from the capes of the Chesapeake to the mouth of the Rio Grande, covering more than 3000 miles of coastline (McPherson 369). Shortly after, she received a letter from President Lincoln's new Secretary of the Navy, Gideon Welles. Because of the blockade order, he informed her, "the fleets needed fresh and large supplies of the Coston Signals." She was all too willing to meet this national need motivated by both patriotism and profit (Coston 82-85).

On August 5, 1861, authorized by an Act of Congress, the U.S. Navy finally acquired her patent. Coston originally asked $40,000. The Senate reduced it to $30,000, and she finally received $20,000.[5] Although the Navy tried to produce the

[4] Parts of this biographical background information has been published and presented in previous works by author.

[5] In the 37th Congress, 1st Session, it was requested by the Secretary of the Navy, Gideon Welles, that "grant of $30,000" should be appropriated for the purchase of the Coston patent. He supports his request with the evidence based on successful trials by boards of naval officers and concluded that these "lights were almost indispensable to the service" (July 26, 1861). The House of Representatives granted the appropriation in the amount of $30,000. Shortly thereafter, the Senate reduced the amount to $20,000. See United States Congressional Publications, 43rd Congress, 2d Session, Report No. 334, March 3, 1875; Coston 90-92. [This note appears

flares at a cost lower than Coston, they were unsuccessful. Consequently, during the Civil War Coston continued to manufacture and supply large quantities of signals at cost to the Navy, which meant selling at a loss due to wartime inflation of production materials and labor costs.

The Official Records of the Union and Confederate Navies in the War of the Rebellion, Series I – Volume 24. Naval Forces on Western Waters From January 1 to May 17, 1863, along with individual ship logs and numerous and various types of correspondence held in the collections of the National Archives, document the extensive use of the signals in blockade operations, strategic battle communication, and rescues on the high sea. In particular, the signals were invaluable in carrying out Lincoln's blockading orders as summarized in the following examples. Additional two brief examples are provided to illustrate the success of the Coston Signal in battle and maritime rescue operations during the Civil War.

Coston claimed that with the Coston Telegraphic Night Signals "nearly all the blockade-runners were caught by their use, as they generally made their runs by night, and the United States navy vessels' gave chase after communication with each other by means of the signals," which was true (Coston 97). Still, risk of capture or worse did not dampen thousands of daring attempts by Confederate runners to bring restricted goods into the South, especially munitions, and to facilitate the export cotton. The transporting of such valuable cargo motivated Union blockaders beyond the call of patriotic duty to capture as many runners as possible, as officers and crews personally shared in the proceeds of captured contraband as judicated through Prize Courts (Soley 44-45). It is estimated that only half of all runners were either captured or destroyed, but this still placed the number of captures upwards of 1,600 (Hooper 108). Coston claimed that many officers "were made rich through prize money by capturing the blockade runners at night, which they did not do in a single instance without the aid of the Coston Signals" (Coston 272; Robington 67-69, 78-80; Schneller 29-31, 115-117).[6]

in my published article, "Martha Coston: A Woman, a War, and a Signal to the World" (2001) note 19. This text appears in my article, "The Use of Coston Flares by the U.S. Navy in Civil War Blockade Operations: A Powerful Auxiliary of Incalculable Value," forthcoming in *Minerva: Women and War*, Spring 2009].

[6] Typically, the government kept half of the money. The commander of the regional blockading squadron received five per cent, and the local squadron commodore received one per cent. The remaining amount was divided into 20 equal shares. Those portions were divided as follows: three to the captain of the capturing ship; ten shares to the officers and midshipmen; remaining seven divided among the enlisted men Some got rich, including high ranking officers. Of particular interest, Admiral David D. Porter, commanding officer of the attack on Fort Fisher and one who was very familiar with the use of the Coston flare and code, reportedly collected over $91,000 in prize money (Simons et al 91; Schneller 29-31,115-117); See also Madeline Russell Robinton, *An Introduction to the Papers of the New York Prize Court*, 1861-1865, 27-29;

The Coston Signal also proved invaluable in maritime rescues during the Civil War. The famous sinking of the iron clad *Monitor* in a storm off the coast of Cape Hatteras on December 30, 1862, dramatically highlights the value of the Coston Signals. The men onboard thought that the "peril was so great that it seemed as if no human power" could save them. The wind had swelled to tornado force and eventually the waves were so high that they rolled completely over the iron turret. From the abstract log of the rescue vessel, the USS *Rhode Island*, it was recorded that a Coston was burned every half hour" from midnight to 4:00 a.m. in an effort to maintain contact and coordinate rescue operations with those still trapped on the *Monitor*. By all accounts, it was a harrowing experience. The official log records that 47 were eventually rescued with the loss of 28 men.

Coston's account collaborates some of dramatic details found in *Harper's Weekly* of January 24, 1863, and later that same year in the October issue of Harper's *New Monthly Magazine,* as well as in survivor narratives. Coston felt "particularly thankful" that her signals played an important part in saving lives as well as capturing the enemy (Coston 102-105; USS *Rhode Island* 353).[7] The effective use of the signals in rescue operations during the war predicted their post-war application in the U.S. Life Saving Service.

In addition to blockade operations and sea rescues, the Coston Signal had another equally significant application in strategic and logistic battle communication. Like the sinking of the *Monitor*, the Battle of Fort Fisher on January 13-15, 1865, provides a dramatic example of just how valuable and extensive the use of the Coston Signal became by the war's end. Rear Admiral David D. Porter, Commander of the North Atlantic Squadron, commanded 60 warships and coordinated with an infantry force of 8,000 under General Alfred Terry (McPherson 820). Admiral Porter issued General Order No. 78, on January 2, 1865, issuing formation orders of the first and second battle lines in this massive attack. The ship logs of the leading three vessels in Battle Line No. 2, the *Minnesota,* the *Colorado*, and the *Wabash,* suggest the complicated strategic goals of the operation and the important part that Coston signals played in the capturing of Fort Fisher.

Many years later, Coston received a letter from Admiral Porter, which she included in her autobiography. The Admiral shared his heartfelt gratitude and memories in the concluding paragraph of his letter:

53-89; and Rebecca Livingston, "Civil War Cat-and-Mouse Game: Research Blockade Runners at the National Archives," *Prologue* 31.3 (Fall 1999).

7 See also: "First Cruise of the Monitor Passiac," *Harper's New Monthly Magazine* 27.161 (October 1863): 577-599; and Francis Banister Butts, "My First Cruise at Sea and the Loss of the Iron-Clad Monitor." *Personal Narratives of the Battles of the Rebellion* Being Papers Read before the Rhode Island Soldiers and Sailors Historical Society (No 4. 1878).

> I shall never forget the beautiful sight presented at ten o'clock at night when Fort Fisher fell. I was determined to be a little extravagant on that occasion, and telegraphed by the signals to all creation that the great fort had fallen and the last entrance to the Southern coast was closed. The order was given to send up rockets without stint and to burn the Coston Signals at all the yard-arms, mast-heads, along the bulwarks, and wherever on shipboard a light could show. The sea and shore were illuminated with a splendor seldom equalled [sic]. . .
>
> What could there be more beautiful than the Coston signals on that occasion, and what more could I say of them? (100)

Although Coston received many such accolades testifying that the signal proved successful beyond a doubt during the Civil War, she felt Congress and other powerful men in the U.S. Navy resented her success and charged that "the chivalry of men towards women" vanished "like dew before the summer sun when one of us comes into competition with the manly sex" (272). Her struggle for financial success continued through the post-war decade.

Foreign Navies

Coston's international experience presents a fascinating glimpse of an inventor who was simultaneously a woman of her time and woman well ahead of her time. Her strategies for success document some of the obstacles, challenges, and accomplishments she experienced while engaged in the business of inventing. In particular, her efforts to market her signal and code system in Europe document an ironic juxtaposition of strict adherence to feminine Victorian protocol and aggressive masculine business strategies.

Her ambitious business plan took nearly 15 years to successfully accomplish during which time she took out patents for her signal flare in England, France, Holland, Austria, Denmark, Italy, and Sweden (55, 60).

Like many of her male inventor counterparts, her experiences with various governments reveal that she met with "opposition, apathy, and prejudice" (Rossman 160).[8] But unlike her male counterparts, she met these challenges with a

[8] In his book, *Industrial Creativity, the Psychology of the Inventor*, Joseph Rossman profiles only male inventors and does not explore gender differences in any context other than to use the gender neutral reference to "all." In Chapter XI, "Obstacles and Pitfalls of Inventors," originally printed in the *Journal of the Patent Office*, 1930, Rossman states that the greatest obstacles an inventor meets are in the "external environment" such as "economic conditions, the prejudices of people, the dishonesty of some promoters, the problems of manufacturing and selling" (161). Although Rossman consistently uses the pronoun "he" throughout his work, Coston's experiences suggest that the challenges were not gendered in nature, but rather

high degree of success by manipulating gendered standards of appropriate behavior to her best business advantage. She depended on a carefully scripted image that exuded proper decorum, appearance, and socially identifiable American characteristics of the gentle well-bred woman. She was well aware of accepted standards that defined a "lady": motherhood, purity, piety, and a sense of fragility that made her appear as if she needed protection and guidance. Her strategies for success in the scientific and business arena depended on feminine entitlement and masculine prerogative.

In 1854, social theorist George Fitzhugh, wrote, "Women, like children, have but one right, and that is the right to protection" (qtd. in Riley 67). Coston's experience demonstrates that she clearly understood this perception, which she used in clever ways whenever it suited her business purpose. When it did not suit her purpose, she used the exact opposite strategy by presenting herself as a strong, independent, and shrewd businesswoman, a woman bent on making the best profit possible regardless of how others viewed it. She shared an American cultural identity with the idea of the self-made man and Yankee ingenuity, with scant regard to the reality that this ideology excluded women. At times, publicly Coston feigned a lack of technological knowledge and business acumen, but her curious intuition, relentless persistence, and carefully forged professional networks belie this public persona.

In August of 1859, Coston landed in Liverpool in route to London. Her first experience foreshadowed how gendered strategies would play a key role in her success. She arrived in Liverpool late, so she could not go on to London until the next day. She left her bags at the docks, accepting the offer by a fellow-passenger to pick them up for her the next morning. Coston wrote that he found the bags easily, but was "disconcerted by finding with it a box of my signals, which had been placed in the care of the purser of the steamer to be stowed with the ship's fireworks. Through some mistake, perhaps because my name was plainly upon it, it had been put with the passengers' baggage." This mistake was a "dreadful violation of British law" and required fast words and quick money to overcome. Coston felt no compulsion to reveal the true contents of her bags to authorities and had no qualms in allowing a bribe by her fellow male passenger to fix the situation satisfactorily. In fact, she was grateful for the entire outcome and learned a valuable lesson that she would employ on later occasions. She could, in effect, hide behind the guise of a proper woman when necessary without suspicion. She understood a lady's baggage was relatively trustworthy in transporting nothing more scandalous than fancy undergarments. It was not suspected that she was transporting contraband, particularly of a military nature. This subterfuge and assumption about the

were part of the typical process and business of inventing. It was Coston's responses and solutions that reflect a distinct gendered approach.

contents of a lady's luggage served her well when she later traveled between France and Germany during a period of military conflict (Coston 58).

In the spring of 1866, the French Minister of Marine offered Coston $8,000 for the patent. With reluctance, she accepted. She considered this a gross underpayment, but reconciled that it was better than no sale. While not terribly significant from a financial perspective, the French sale gave her work momentum. By 1867 an Official Bulletin of the French Navy was published, which detailed the testing and adoption of "Coston's Telegraphic Night Signals" (*Coston's Telegraphic Night Signals* 10-11). Her persistence, network development, and business intuition all worked together to propel her to further successes in Europe.

The year 1867 found Coston's goals focused on Italy. She gained valuable business experience in her relationships with England and France, and she continued to pay close attention to all things proper for an American lady. Her first foray into Italian business was to boldly announce her arrival by card and letter to ranking Italian Naval commanders that introduced both the invention and the inventor (157, 153). In addition to business adventures, Coston's personal life took several interesting turns while in Italy.[9] Although she had strong feelings about marriage, specifically about not marrying, she became engaged to an Italian count. Her account reads like a Victorian romance with the Count dying a tragic and suspicious death before they were married. This experience provided context for her to express her view on marriage. [10] Like most other professional American women inventors of the 19th century, Coston remained single for the rest of her life.[11]

[9] She had the opportunity of being presented to the Pope. A protestant herself, she described the ceremony as "a wholesale way of doing things." This experience contrasted with her reaction when presented to the Queen of England several years earlier. She described that experience as brimming with pomp and ceremony, a Cinderella-like experience. Like her business acumen, her impressions of European nobility matured (119-121; 164). When introduced into Italian society, she was not so much impressed with Italian society itself, but rather how she was idealized and treated with profound reverence (167). Her success as a businesswoman added confidence to her sense of entitlement.

[10] Up to this point, Coston made it clear on several occasions that she was determined to remain single. Marriage rather than employment was the preferred form of economic security for a lady, but Coston tenaciously prided her independence. But early during her trip to Italy she met and fell in love with Count Piccolomini, a member of a prominent Italian family. After a brief courtship, they were engaged. Tragically and suspiciously, the Count became gravely ill shortly after the announcement of their engagement. Within a matter of days, he died. Despite numerous investigative attempts, Coston never received confirmation as to the cause of his death, but suspicion weighed heavily with her in the fact that the Count was a wealthy man and was the last lineal heir of the Piccolomini family. Dying unmarried, his immense fortune went to his relatives. If married, it would have gone to Coston, who was not only a foreigner, but a Protestant as well (173-175). Upon later reflection, Coston's ideas about marriage had not been tempered by her near capitulation. She wrote, "During my long sojourn abroad I met many American women bearing the titles of baroness, countess, my lady, and occasionally duchess and princess. As a rule their marriages had been made on the usual basis of exchange,

Coston continued over the next three years to make her way through war-torn Europe attending to business and solidifying the credibility and value of the signals and code system. In 1872, Holland, officially notified Coston that her signals proved better than the night signals they were using in the Dutch Navy, hence they were officially adopted (*Coston's Telegraphic Night Signals* 12).

At this point in time, Coston once again turned her attention back to business in the United States, where she established a strong relationship with the U.S. Life Saving Service, private yachting clubs, and American and foreign shipping and merchant companies. Her European experiences reflect the complexities inherent in the business of invention and the gendered strategies employed in meeting her unwavering goal: success.

U.S. Life Saving Service

By the 1870s, Coston's sons were both grown young men, and William F. continued to make the family business his life's work. He took the lead in working with the USLSS and devised a "plan to provide distinguishing signals for different lines of steamers and other craft, yachts, etc." He was awarded four patents in his own name, all related to the original pyrotechnic signal.[12] Coston's other son, Harry H. Coston, (sometimes referred to as Henry or simply as H.H.) was a professional soldier in the Marine Corps, but he remained involved with the company, providing ideas for improvements and business support (Coston 292). Coston herself, however, remained at the helm of the company as "the principle party interested in the business" and held the reputation that business was mainly carried on under her direction (Dun Credit Report, February, 1887, 1433).[13]

— gold for a name; and as a rule, naturally enough, the marriages differed only in degrees of misery" (195).

[11] Professional women inventors did not fit the typical mold of an ideal woman, but by the end of the 19th century an early profile emerged among American professional women inventors. "She was generally single and without children or widowed with dependent children. She invented for her livelihood. Her inventions were often mechanical, mostly commercial, and achieved a standard of technical sophistication with manufacturing potential. She was an oddity and made to feel deviant for pursuing unnatural priorities even if her inventions were domestic" (Pilato 109). Coston's carefully attention to feminine ideals functioned progressively for her as she established herself in a masculine scientific and business world.

[12] William F. Coston Patents: Patent No. 237,092, Feb 1, 1881: Pyrotechnic Signal; Patent No. 570,458, Nov 3, 1896: Pyrotechnic Signal Holder; Patent No. 658,498, Sept 25, 1900: Compartment Box; Patent No. 674,400, May 21, 190: Signal Holder.

[13] Dun Credit Reports for William F. Coston reveal interesting details about the reputation of Martha J. Coston. Several references are made to William's "fair ability and financial responsibility" in comparison to his mother's reputation as the principal company director. The trade considered him "honest" but did not think that he had "much means and have always supposed

Included in the *Annual Report* for the year ending June 30, 1883, is a letter from William F. Coston to General Superintendent Kimball urging, "a night signal be established for the purpose of distinguishing a Life-Saving Service station. Coston hoped to have Kimball's "early and favorable answer" so he could publish it as part of the 1882's night signal chart. The purpose of this signal was to aid a vessel in determining where they were located along an unfamiliar coast in relationship to a particular station. Said signal would provide all required information needed for a safe passage and thereby, Coston argued, it would be a valuable "means of preventing wrecks" and would prove to be of the utmost importance in avoiding risk for both vessels and rescuers alike (501). This push for a new specific signal code adoption attempted to further diversify the application of Coston signals and codes, because as competitors entered the market as noted in the same year of the *Annual Report* (403). While his mother maintained a leading role in company management, William worked diligently to keep the USLSS satisfied with Coston Signal Company products, especially in light of some published unfavorable comparisons that were made after the USLSS's testing of the Coston signal with one of its competitor's, the Jackson Signal (458-460).[14]

However, the success of the Coston Signal and its adoption by the USLSS continued strong and became a standard of the industry over the next two decades. By the turn of the century, each of the *Annual Reports* included hundreds of instances documenting the use of the Coston. For example, the 1901 *Annual Report* under "Vessels Warned From Danger" stated that "231 vessels were warned away from dangerous places by the signals of the patrolmen" and that, "210 of the instances occurred in the night, frequently during fog, rain storms, or snowstorms, and a large proportion of them in freezing weather" (181). The closing comments under this section reiterated the important service performed by the beach patrol and aided by Coston Signals.

> No part of the work performed by the Life-saving Service is more important or more entitled to the gratitude of the seafaring and commercial world than the patrol, which guards the entire coast within the limits of the Service during all hours of the night, from sunset to dawn, and during the daytime in thick and stormy weather. (181)

The Coston Signal was a ubiquitous piece of equipment carried by all patrolmen by this time.

his mother" was in charge of the business. Sometimes William received credit and other times he did not. No records were found requesting loans by Martha J. Coston (*Dun Credit Report*).

[14] William also took the lead in selling the signal to railroad companies who also had steamship lines. The adoption of the Coston Night Signal by the Central Pacific Railroad Company in 1882 was announced in the *NY Times*, September 21, 1882, 8.

The Coston Supply Company

When Coston's autobiography was published in 1886, there were "sixty or seventy different interests using the distinguishing signals" (Coston 305, 307). Nearly 100 years later, in 1985, the Coston Supply Co. was still listed as doing business in New York City.[15] Although details about her work as an inventor are only just now coming to light, it is clear that the Coston Supply Co. remained a solvent company throughout most, if not all, of the 20th century.

Coston's husband, Benjamin Franklin Coston, was 19 years old when he started the original company in 1840. After his death in 1848, his business associates claimed that all the "ready capital" had been depleted. The company was bankrupted for all practical purposes. His widow quickly realized that she needed a better understanding of business as she had not made a practice of demanding accurate accounts of her husband's business associates. Too late, she realized it was through her "own ignorance and duplicity of others, trusting too much to an improvident relative who misplaced" her money, that she was alone in the world without financial resources (37). Years later, in true Victorian style, she confessed,

> It would consume too much space, and weary my readers, for me to go into all the particulars of my efforts to perfect my husband's ideas. The men I employed and dismissed, the experiments I made myself, the frauds that were practiced upon me, almost disheartened me; but despair I would not, and eagerly I treasured up each little step that was made in the right direction, the hints of naval officers, and the opinions of the different boards that gave the signals a trial. (42-43)

By 1859, Coston not only was awarded a patent, but the Coston Signal Company became operational under her sole management. Although she was still subject to financial disappointment due to wartime inflation, Congressional deceit, and gender discrimination, she built her company into a financial success, and her invention was a viable technical achievement well into the 20th century.

When she penned the final lines of her autobiography, she acknowledged that she had done her "woman's share of the fighting" and felt satisfied for all of her efforts. The last paragraph of her autobiography concludes with commentary on the battle that preoccupied her in 1886. She demanded that the government erase the "name of Very from the Coston Signals" which they were presently using. The

[15] *The Directory of Directors in The City of New York* lists the Coston Supply Company consistently from the years 1911-1985. The company name was initially listed as the Coston Signal Co. and changed in 1927 to Coston Supply Co. It is unknown at the time of this writing, whether or not the company survived longer or is still in existence today under another name.

Navy department submitted to the Superintendent Kimball of the USLSS a code titled "Coston or Very" signal Code." When Coston learned about this, she was "indignant that this lieutenant [Edward Wilson Very, 1847-1910] should presume to place his name beside mine." She tried different tacks for the removal of the Very name, even repeatedly requesting an audience with President Grant. For a time, her efforts proved in vain. However, she was unwilling to "suffer in silence" for long. She successfully applied to the Secretary of the Treasury for redress and sarcastically noted that in this instance the Navy did not think it wrong "to tamper with the commercial code" (Coston 296, 313; *Signal Pistols*). Even though successful, the usurpation irked her for the rest of her life.[16]

Her persistent fight for identity as inventor fueled a final spitfire of fighting spirit in the next to last paragraph of her 9-page "Last Will & Testament." She left final instructions regarding a lawsuit that was before the United States Court of Claims for "recompense for the use of by the United States Navy" of her invention of "An Aerial Night Signal" which she said was legally transferred to her from her son Henry.[17] She was nearly 75 years old at this time, and her maturity was not only measured in years, but in her wide range of scientific knowledge and hard won business success. Unlike the vulnerable and ignorant young woman of the 1850s, she exhibited "no fear or hesitation" in later life when it came to fighting for her due recognition and reward (333). By the end of century both the inventor and her invention received numerous awards and prestigious international and national recognition.[18]

[16] "Edward Wilson Very invented a "new and useful improvement in Pyrotechnic signal cartridges . . . ." US Patent 190263 dated 1 May 1877. Very is predated by one Benjamin Franklin Coston, but Coston's gun retained the flare & was really only an ignition device and the "gun" was waved at arms length. Very's gun discharged the flare in the way we know today. In a quirk of history, Coston's son actually invented the first aerial flare launching cartridge, but his mother, jealous of the invention, lobbied against it being accepted by the Navy. Therefore Very got the credit for inventing the pistol & cartridge. Very's invention was accepted in 1882 by the US Navy & by 1900 Very pistols & cartridges were in use throughout the world." The reference to Coston's jealously of her son's inventive ambition is referenced in popular sources, but her opposition has yet to be found in government documents (*Signal Pistols*. <http://www.diggerhistory.info/pages-weapons/signal-pistols.htm>; Andrew Lustyik. "History of Military Pyrotechnics, Part II." *The Gun Report* May, 1968: 23-24).

[17] In 1877 Henry (also referred to as Harry) filed a patent in 1877 for Aerial Signals, which "shot so the light ascended higher up in the air." The patent did not have much success, which some contribute to his mother's working against its success (Frank Russell. "Early U.S. Martial Signal Pistols." *The Gun Report* August, 1970: 14).

[18] Coston Signals received the following awards: 1873, Vienna, Austria, Diploma and Medal; 1875, Chili, Santiago de Chili, Diploma and Award; 1876, U. S, Centennial at Philadelphia, Medal and Diploma; 1893, World's Colombia Exposition at Chicago. End page of *A Signal Success*.

The company continued to operate after her death in 1904 in New York City under the name of the Coston Signal Company until 1927, when the name changed slightly to the Coston Supply Company (*Directory of Directors in the City of New York* 179; 187). Intriguing bits of information surface from time to time in unlikely sources suggesting the continued growth of the company long after Coston's death. A 1918 complimentary notepaper pad, designed as a free give-away advertisement, includes an image of the handsome office building on Water Street where the company was located. It also includes a list of "Coston Specialties" with a brief discussion of how their "Davit Turning Out Gear" had "solved the problem of malfunctioning boat davits in cold weather in a most satisfactory manner. The Coston Night Signal was still featured as a prominent product, complete with illustration. This small advertising promotional piece suggests the diversified growth of the company (Coston Supply Company Artifacts. Complimentary Note Paper & Advertisement Booklet 1918).

What is certain is that the business grew under Coston's leadership and prospered for more than 125 years. In addition to use by the U.S. Navy, foreign governments, and the USLSS, the Coston Signal was used by railroads, merchant vessels, yachting clubs, public ferry services, early aviation, and sometime in unusual ways, such as part of expedition gear by explorers. Coston was particularly happy with this application by Lieutenant Schwatka's 1880 Arctic expedition, when they were used to warn away wolves. Coston learned of this while listening to a public Geographical Society lecture given by Schwatka. After the lecture, she remarked to him that he had used the signals to carry out her original idea. She said her "principal object in perfecting the invention was to keep the wolf from my own door" (Coston 308, 328-331; Russell 15; Gilder 61-62).

Conclusion

This principal objective was significantly reflected in the company's original and only trademark, which was a life ring. Inside the life ring was a hand holding an ignited signal flare circled by the Latin words "In hoc lumine pes mea," which loosely translated means "in this light is my hope." Coston's own hopes for success were realized, as well as thousands of lives that were saved through the use of the Coston Signal.

During her life, she traversed many geographic and cultural boundaries. Some were demarcated by nationality, others by distinctive social and gender constructions. Her achievements challenged the popular and persistent convention that women did not invent anything of significance, and her career represents an early prototype of a woman inventor. Like other women whose professional identity can be categorized as inventor, she remained single, received patents, struggled for educational access and scientific advancement, established a business, fought for legal and financial entitlement, and insisted on maintaining company control.[19] As the French correctly noted, her position was a peculiar one, one that peculiarly led her to success in the business of invention.

She forged her identity as an inventor during an era when an inventor was implicitly male and often experienced gender discrimination when she competed with "small minded" men who begrudged a woman her success (Coston 272). The significant differences in the measure of success between the invention verses the inventor lay in the perception of those who write the history and those who construct criteria for significance. It is not surprising that Coston wrote her own story in light of her obsession with success and identity as an inventor. There had never before been a celebrated woman inventor, one who ranked recognition in the American canon of important inventors.

The 20th century was not forth coming with recognition for women inventor's accomplishments. In the early years of the 21st century, Coston's experiences as a 19th woman inventor can be viewed as an early shifting paradigm not only in the gendered idea of inventor identity, but in the reality of significant technological accomplishments.

Martha J. Coston's contributions as an inventor are significant technically, financially, and socially because her life's work presents an illuminated pattern of success. In addition to a pure white, a vivid red, and brilliant green color, her signals not only flashed clearly across a dark sky, but signaled to the world that gender was merely a mitigating circumstance and not a determination for failure, relegation, or insignificance in the history of technology.

Works Cited

Annual Report of the Operations of the United States Life-Saving Service. Washington, D.C.: Government Printing Office: Year Ending June 30, 1883.

Annual Report of the United States Life Saving Service. Washington, D.C.: Government Printing Office: Year Ending June 30, 1883.

Annual Report of the United States Life-Saving Service By United States Life-Saving Service 1901 [retr 7-7-07] <http://books.google.com/books?id=q4ADAAAAYAAJ&ots=nzOhog8fzU&dq=annual%20report%20united

[19] Examples of professional women inventors: Margaret Knight, Harriet Ruth Tracy, Clarrissa Britain, Helen Augusta Blanchard, Amanda Theodosia Jones, Maria Beasly, and Mary Carpenter all fall within the category of professional inventor (See Pilato, ch. 5, "National Reconstruction versus Gender Construction" 107 139).

%20states%20life%20saving%20service&pg=PA181&ci=97,316,837,1068&source=bookclip>.

Bradlee, Francis B. *Blockade Running During the Civil War and the Effect of Land and Water Transportation on the Confederacy.* Philadelphia, PA: Porcupine Press, 1974.

Brown, J. Willard. *The Signal Corps, U.S.A. in the War of the Rebellion.* Boston, MA: U.S. Veteran Signal Corps Association, 1896.

Butts, Francis Banister. "My First Cruise at Sea and the Loss of the Ironclad Monitor." *Personal narratives of the Battles of the Rebellion, being papers read before the Rhode Island Soldiers and Sailors Historical Society.* No. 4. Providence: S.S. Rider, 1878.

Carpenter, Frank G. "Uncle Sam's Life Savers." *The Popular Science Monthly* 1 (1894): 346-353.

Cochran, Hamilton. *Blockade Runners of the Confederacy.* New York: Bobbs-Merrill Company, Inc., 1958.

Coston, Martha J. "Last Will and Testament." 21 January 1903. (Washington, D.C.: 21 January 1904).

---. *A Signal Success: The Work and Travels of Mrs. Martha Coston.* Philadelphia, PA: J.B. Lippincott Company, 1886.

Coston Supply Company Artifacts. Author's collection: Congressional Petition for Compensation re: Cannon Percussion Primer (1874), Complimentary Note Paper & Advertisement Booklet (1918); Coston Compass n.d.; Flare Canister & Flares n.d. Coston Supply Company product catalogue pre-WWII.

Coston's Telegraphic Night Signals Patented in the United States and Europe, and Adopted by the Governments of the United States, France, Italy, Denmark, Holland, Hayti, and the New-York, Brooklyn, and Eastern Yacht Clubs; used in the U.S. Revenue, Marine, Life-saving and Light-house Services; Diploma of Merit, Vienna, 1873. New York: S.W. Green, 1880.

Directory of Directors in the City of New York. Directory of Directors Company 1913-1985.

Dun Credit Reports. Coston WmF. Vol. 448. 1433; 1500 (1880s-1890).

"First Cruise of the Monitor Passiac." *Harper's New Monthly Magazine* Vol. 27.161 (October 1863): 577-599.

Gilder, William H. *Schwatka's Search Sledging in the Arctic in Quest of the Franklin Records.* New York: Charles Scribner's Sons, 1881.

Hampden, Augustus Charles Hobart. *Never Caught Personal Adventures Connected with Twelve Successful Trips in Blockade-Running During the American Civil War.* Carolina Beach, NC: The Blockade Runner Museum, 1967.

Hayes, John D. (ed.) *Samuel Francis Du Pont: A Selection From His Civil War Letters. Volume I: The Mission: 1860-1862.* Ithaca, NY: Cornell University Press, 1969.

Hoehling, Adolph A. *Damn the Torpedoes! Naval Incidents of the Civil War.* New York: Gramercy Books, 1989.

Hunt, O.E. (ed.) *The Photographic History of the Civil War. Volume 5: Forts and Artillery.* New York: Thomas Yoseloff, 1957.

King, Irving. *The Coast Guard Expands 1865-1915 New Roles, New Frontiers.* Annapolis, MA: Naval Institute Press, 1996.

King, William C. and William P. Derby. *Camp-Fire Sketches and Battlefield Echoes 61-65.* Springfield, MA: King, Richardson & Co, 1886.

Laycock, Thomas F. *North Atlantic Blockading.* Lithograph. New York: Endicott & Co., 1865. 1 July 2007 <http://americancivilwar.com/statepic/nc/nc015.html>.

Livingston, Rebecca. "Civil War Cat-and-Mouse Games: Research Blockade-Runners at the National Archives." *Prologue* 31.3 (Fall 1999): 1-9.

Lustyik, Andrew F. "A History of Military Pyrotechnics, Part I." *The Gun Report* 4 (1968): 8-18.

---. "A History of Military Pyrotechnics, Part II." *The Gun Report* 5 (1968): 16-29.

"Marine Intelligence Miscellaneous." *New York Times* 21 September 1882: 8.

"Mr. Coston's Wife Sues." *New York Times* 14 July 1901: 3.

National Archives. Record Group 24. "Log of Ships & Stations." *Wabash* 13 January 1865.

---. Record Group 45. "Subject File, U.S. Navy 1775-1910." Box 161 Signals General.

---. Record Group 45. "Subject File, U.S. Navy 1775-1910." Box 162, Envelope: Gulf Blockade Squadron, Distinguishing Pennants and Lights.

---. Record Group 45. "Subject File, U.S. Navy 1775-1910." Box 186, Folder: December 1864-1865, Bombardment of Fort Fisher.

Noble, Dennis L. *That Others Might Live: The U.S. Life-Saving Service, 1878-1915.* Annapolis, MA: Naval Institute Press, 1994.

"Petition of Mrs. M. J. Coston to the Congress of the United States for the use of the Inventions of the late Benjamin Franklin Coston; Particularly that known as the Cannon Percussion Primer." Philadelphia: Sherman & Co Printers, 1874. [author's collection]

Pilato, Denise E. *The Retrieval of a Legacy: Nineteenth Century American Women Inventors.* Westport, CT: Praeger, 2000.

Porter, David Dixon. *Incidents and Anecdotes of the Civil War.* New York: D. Appleton and Company, 1891.

Riley, Glenda. *Inventing the American Woman.* Arlington Heights: Harlan Davidson, 1987.

Robinton, Madeline Russell. *An Introduction to the Papers of the New York Prize Courts, 1861-1865.* New York: Columbia University Press, 1945.

Russell, Frank. "Early U.S. Martial Signal Pistols." *The Gun Report* 8 (1970): 10-15.

Schneller, Robert J., Jr. *Under the Blue Pennant or Notes of a Naval Officer*. New York: John Wiley & Sons, 1999.

Shanks, Ralph, Wick York, and Lisa Woo Shanks. *The U.S. Life-Saving Service: Heroes, Rescues, and Architecture of the Early Coast Guard.* Petaluma, CA: Costano Books, 1988.

Signals Pistols. Retrieved 29 August 2008 <http://www.diggerhistory.info/pages-weapons/signal-pistols.htm>.

Simons, Gerald, Henry Woodhead, Herber Quarmby, and Philip Brandt George et al. (eds.) *The Civil War. The Blockade Runners and Raiders*. Alexandria: Time-Life Books, 1983.

Smith, Darrell Hevenor and Fred Wildbur Powell. *The Coast Guard: It's History, Activities and Organization*. Washington, D.C.: The Brookings Institution, 1929.

Soley, James Russell. *The Blockade and the Cruisers*. New York: Charles Scribner's Sons, 1883.

Taylor, Thomas E. *Running the Blockade A Personal Narrative of Adventures, Risks, and Escapes During the American Civil War.* New York: Charles Scribner's Sons, 1896.

United States Congressional Publications. "36th Congress, 2nd Session, Ex. Doc. No. 32." Washington, D.C.: Government Printing Office, 15 Janury 1861.

---. "43rd Congress, 2nd Session, Report No. 334." 3 March 1875.

---. "44th Congress, 1st Session, Report No. 259." 12 April 1876.

United States Naval War Records Office. United States Office of Naval Records and Library. *Official records of the Union and Confederate Navies in the War of the Rebellion. USS Rhode Island.* Abstract log. 29 December 1862. 9 April 2007 <http://cdl.library.cornell.edu/cgi-bin/moa/pageviewer?frames=1&cite=http%3A%2F%2Fcdl.library.cornell.edu%2Fcgi-bin%2Fmoa%2Fmoa-cgi%3Fnotisid%3DANU4547-0008&coll=moa&view=50&root=%2Fmoa%2Fofre%2Fofre0008%2F&tif=00381.TIF&pagenum=353>.

---. *Official records of the Union and Confederate Navies in the War of the Rebellion.* / Series I – Volume 21: West Gulf Blockading Squadron (1 January 1864 – 31 December 1864). Washington, D.C.: Gov Printing Office, 1906. 9 April 2007 <http://cdl.library.cornell.edu/cgi-bin/moa/pageviewer?frames=1&coll=moa&view=50&root=%2Fmoa%2Fofre%2Fofre0021%2F&tif=00868.TIF&cite=http%3A%2F%2Fcdl.library.cornell.edu%2Fcgi-bin%2Fmoa%2Fmoa-cgi%3Fnotisid%3DANU4547-0021>.

---. *Official records of the Union and Confederate Navies in the War of the Rebellion.* / Series I – Volume 22: West Gulf Blockading Squadron (1 January 1865 – 31 January 1866); Naval Forces on Western Waters (8 May 1861 – 11 April

1862). Washington, D.C.: Gov Printing Office, 1908. 9 April 2007 <http://cdl.library.cornell.edu/cgi-bin/moa/pageviewer?frames=1&cite=http%3A%2F%2Fcdl.library.cornell.edu%2Fcgi-bin%2Fmoa%2Fmoa-cgi%3Fnotisid%3DANU4547-0022&coll=moa&view=50&root=%2Fmoa%2Fofre%2Fofre0022%2F&tif=00184.TIF&pagenum=161>.

---. *Official Records of the Union and Confederate Navies in the War of the Rebellion,* Series I – Volume 24. Naval Forces on Western Waters from January 1 to May 17, 1863. Washington, D.C.: Government Printing Office, 1911.

---. *Official records of the Union and Confederate Navies in the War of the Rebellion.* / Series Volume 27: Naval Forces on Western Waters (1 January 1865 – 6 September 1865); Supply Vessels (1 January 1865 – 6 September 1865). Washington, D.C.: Gov Printing Office, 1917. 9 April 2007 <http://cdl.library. cornell.edu/cgi-bin/moa/pageviewer?frames=1&coll=moa&view=50&root=%2Fmoa%2Fofre%2Fofre0027%2F&tif=00738.TIF&cite=http%3A%2F%2Fcdl.library.cornell.edu%2Fcgi-bin%2Fmoa%2Fmoa-cgi%3Fnotisid%3DANU4547-0027>.

---. *Official records of the Union and Confederate Navies in the War of the Rebellion.* / Series I – Volume 8: North Atlantic Blockading Squadron (5 September 1862 – 4 May 1863). Washington, D.C.: Gov Printing Office, 1899. 9 April 2007 <http://cdl.library.cornell.edu/cgi-bin/moa/pageviewer?frames=1&coll=moa&view=50&root=%2Fmoa%2Fofre%2Fofre0008%2F&tif=00384.TIF&cite=http%3A%2F%2Fcdl.library.cornell.edu%2Fcgi-bin%2Fmoa%2Fmoa-cgi%3Fnotisid%3DANU4547-0008>.

United States Patent Office. *Decisions of the Commissioner of Patents for the Year 1898.* Washington, D.C.: Government Printing Office, 1898.

---. *Women Inventors to Whom Patents have Been Granted by the United States Government October 1, 1892 to March 1, 1895.* Appendix No. 2. Compiled under the direction of the Commissioner of Patents. Washington, D.C.: Government Printing Office, 1895.

---. *Women Inventors to Whom Patents have Been Granted by the United States Government July 1, 1888 to October 1, 1892.* Appendix No. 1. Compiled under the direction of the Commissioner of Patents. Washington, D.C.: Government Printing Office, 1892.

---. *Women Inventors to Whom Patents Have Been Granted by the United States Government 1790 to July 1, 1888.* Compiled under the direction of the Commissioner of Patents. Washington, D.C.: Government Printing Office, 1888.

---. *Letters Patent No. 115,935* (13 June 1871).

---. *Letters Patent No. 23,536* (5 April 1859).

---. *Official Gazette.* 25 March 1873: 325.

USS Monitor (1862-1862) – Selected Views. Naval Historical Center <http://www. history.navy.mil/photos/sh-usn/usnsh m/monitor.htm>.

Wilkinson, John. *The Narrative of a Blockade Runner.* New York: Sheldon and Co., 1977.

Whiting, John D. "Knights of the Wave: the Story of Rescue at Sea." *The Mentor* July 1925: 1-17.

Disciplining Technopolitics: Physics, Computing, and the "Star Wars" Debate

Rebecca Slayton

The cultures of science and engineering are many. If this has become a truism within Science and Technology Studies (STS), it is a testament to an impressive body of scholarship built upon Thomas Kuhn's revolutionary work. For example, Karin Knorr-Cetina has detailed the distinctive material cultures of experimental particle physics and molecular biology. The distinctive goals of science and engineering shape the cultures of testing; where science privileges representation, engineering privileges reproducibility (Downer). As transnational studies by scholars such as Sharon Traweek, Gary Downey, and Juan Lucena have shown, science and engineering cultures vary with place. So too, do they vary in time. Ironically, even as C.P. Snow mused about the gap between two monolithic cultures—science and the humanities—Cold War politics fostered the proliferation of specialization, and new science-engineering hybrids (Galison). And as David Kaiser has recently shown, even the abstract tools of theory are tied to historically contingent practices of training and research. To describe these findings is to merely skim the surface of a much deeper and wider ocean of work.

But while we know of myriad ways that public politics shapes the cultures of science and engineering, we know little of how distinctive disciplinary cultures shape public politics. At a glance, this seems odd. Whether we speak of "mutual shaping," or "co-production," STS scholars recognize that causation runs both ways: scientific and public cultures influence one another. But traditional methods for analyzing scientists' influence on public politics have tended to elide any examination of disciplinary specificity. Analyses have focused on particular advisory institutions, such as the U.S. President's Science Advisory Committee (Herken), the National Academy of Sciences (Hilgartner), advisory panels for the Environmental Protection Agency (Jasanoff), or the International Panel on Climate Change (Sluijs et al.). Alternately, Steve Epstein, Kelly Moore, and others have focused on less institutionalized activist groups. All such groups are, of necessity, interdisciplinary.

This essay confronts the relationship between disciplinary and public cultures by taking disciplines as a unit of analysis. In particular, we will see how two distinctive disciplines—physics and computing—shaped public debate about President Ronald Reagan's 1983 proposal to develop missile defenses that would render nuclear weapons "impotent and obsolete." Formally known as the Strategic Defense Initiative (SDI), but popularly known as "Star Wars," Reagan's vision was riddled with uncertainties that were simultaneously technological and political.

How did scientific disciplines shape the ways that experts grappled with these uncertainties? And how did experts translate highly technical analyses into arguments that were compelling in the turbulent and changing world of public politics? The first question highlights the multiplicity of scientific cultures, while the second turns our attention to the specificity of political culture (Jasanoff "Designs on Nature").

Here I will focus on a key practice at the crux of these two questions, what I will call *disciplined projection*. Confronting an uncertain future, intertwining technology and politics, scientists use generally accepted, mathematical rules of disciplines to project an element of certainty into the future. This is pragmatically useful, allowing experts to analyze a problem in quantitative, predictive terms. Importantly, different disciplines highlight distinctive aspects of the same problem. Just as a gestalt shift changes a rabbit into a duck, the emergence of new discipline can radically alter scientists' perception of a complex technological problem. Disciplined projections are also useful because they carry a kind of social authority, what Ted Porter has termed "disciplinary objectivity."

But ultimately authority emerges not only from experts' discourse, but from the political culture which gives it meaning and relevance. As we will see, two specific aspects of U.S. political culture were crucial for the SDI debate.

First, fears of technocracy have prompted the maxim that scientists are to represent but not to intervene in politics. As Sheila Jasanoff and other scholars have shown, efforts to "speak truth to power" fall short; boundaries between "science" and "policy" are contingent and negotiated (Jasanoff *The Fifth Branch;* Jasanoff "Science, Politics, and the Renegotiation of Expertise"). Nonetheless, disciplines that profess to represent rather to intervene, and to speak for nature rather than artifacts, can more easily mind such boundaries. For example, the politics of theoretical physics, however real, are much less visible than those of civil engineering.

Second, as Thomas Hughes, Sheila Jasanoff, and others have argued, the United States history has a long tradition of enthusiasm for new technology. Initially a resource-rich and labor-scarce land, the U.S. built a national identity upon the wedding of technology and free market capitalism. American faith in technological progress has been institutionalized in modern corporations. As a result, critics of new technology face pressure not only to show that a technological proposal is probably flawed, but that it is provably flawed. As we will see, this was especially true of the "Star Wars" missile defense proposal.

1. Star Wars: Technological Progress vs. Countermeasures

In a nationwide televised address on March 23, 1983, President Ronald Reagan put a fresh face on his increasingly unpopular defense policies. After emphasizing that the Soviet Union nuclear threat was growing and required increased spending on U.S. nuclear weapons, he changed his tone, expressing his deep and growing conviction that "the human spirit must be capable of rising above dealing with other nations and human beings by threatening their existence." He soon connected his vision to deeply American narratives of technological progress:

> Let me share with you a vision of the future which offers hope. It is that we embark on a program to counter the awesome Soviet missile threat with measures that are defensive. Let us turn to the very strengths in technology that spawned our great industrial base and that have given us the quality of life we enjoy today.
>
> What if free people could live secure in the knowledge that their security did not rest upon the threat of instant U.S. retaliation to deter a Soviet attack, that we could intercept and destroy strategic ballistic missiles before they reached our own soil or that of our allies?

Reagan suggested that technical progress could shift U.S. nuclear strategy away from offense to defense, and called upon "the scientific community . . . to turn their great talents now to the cause of mankind and world peace, to give us the means of rendering these nuclear weapons impotent and obsolete." When journalists and policymakers learned that this visionary missile defense would include space based weapons such as lasers, they promptly dubbed Reagan's proposal "Star Wars" (Cannon "President Seeks Futuristic Defense").

Scientists and engineers unanimously agreed that missile defense alone could not achieve Reagan's goal of rendering nuclear weapons impotent and obsolete. In fact, Reagan's commissioned study, conducted by fifty scientists and engineers under National Aeronautics and Space Administration Director James Fletcher, concluded that a 99.9% effective defense was "not technically credible" (Robinson "Study Urges Exploiting Technologies"). As a companion "Security Strategy Study" made clear, this meant that at most, missile defense would enhance deterrence by protecting missile silos, not eliminate nuclear deterrence by protecting people (Hiatt "Limited ABM"). However, the Reagan administration continued to insist that the missile shield would protect people (Hiatt "Reagan's 'Star Wars' Uncertain").

The administration and others in the defense industry focused on a different aspect of the Fletcher panel: enthusiasm for new technology. As presented to Ronald

Reagan and leaked to an enthusiastic defense industry in October of 1983, the Fletcher report recommended that the U.S. spend over 20 billion dollars developing directed energy weapons, optics and steering systems, hit-to-kill vehicles, and other exotic technologies (Robinson "Panel Urges Defense Technology Advances"). A commitment of over 20 billion dollars in the second half of the 1980s would merely provide the basis for a decision to deploy a "tiered" defense system. Such a system would aim to stop missiles in every phase of flight – boosting out of the earth's atmosphere, traveling through space in the midcourse phase, and reentering the atmosphere in the terminal phase (see Figure 1).

Many journalists viewed the Fletcher report's enthusiasm for new technology as a sign that Reagan's vision was achievable, even though the report briefly acknowledged otherwise. Much was at stake in this misunderstanding. The Fletcher report helped justify Reagan's Strategic Defense Initiative (SDI), a massive research, development, and demonstration program. As others have noted, the creation of the SDI Organization (SDIO) within the defense department was controversial, as it swallowed programs belonging to the services and the Advanced Research Projects Agency, but it supported the defense industry and the national weapons laboratories at a rate of up to $4 billion dollars per year during the Reagan years (Fitzgerald; Broad; Pratt).

Furthermore, SDI was a source of significant international tension. Soviet opposition was immediate, ongoing, and widely publicized (Gwertzman). It became most stark in what historian Francis Fitzgerald has called "the most bizarre summit in the history of the Cold War": a meeting between Reagan and Mikhail Gorbachev in Reykjavik, Iceland (Fitzgerald 315). After discussing the possibility of drastically reducing nuclear weapons arsenals, Gorbachev insisted that the United States limit tests of "Star Wars." Reagan refused to accept limits, and the meeting stalled. In a widely publicized breakdown, Reagan and Gorbachev parted ways angrily (Cannon "Reagan-Gorbachev Summit").

At the crux of the political, strategic, and technological debate that swirled around SDI were countermeasures: could and would the Soviet Union nullify any advanced defensive system with clever technology or offensive buildups? The question was hardly new. In 1958, before any superpower deployed intercontinental ballistic missiles, scientists and engineers were inventing decoys, chaff, and other clever ways to deceive and overwhelm any defense an enemy might devise (Barber). Even without such countermeasures, an effective defense required a tremendously complex assemblage of weapons, sensors, and computers to work together, detecting, tracking, and destroying many missiles, with almost no room for error: just one missile would wreak devastation. The system would have no more than thirty minutes of warning, and limited opportunities for testing. With the deliberate unpredictability of countermeasures, the already limited prospect of proving system reliability grew more limited still.

When scientists and engineers studied ways of building an effective missile shield, they thus confronted the deep uncertainties inherent in countermeasures: could and would they foil any defensive system? As the Fletcher report acknowledged, midcourse and terminal defenses were exceedingly easy to overcome with countermeasures, making boost phase defense crucial to success (Robinson "U.S. Strategic Defense"). A boost phase system was also the most futuristic, but the Fletcher report insisted that "long-term" progress held "particular promise for highly effective and robust counters to Soviet countermeasures" (Robinson "Panel Urges Defense").

This vision of technological progress put the onus on dissenting scientists and engineers to project a different future into the Star Wars debate, to persuade policymakers and their publics that missile defenses could not render nuclear weapons "impotent and obsolete." They needed to prove impossibility, or at the very least, improbability. In what follows, we will see how the disciplines of physics and computing shaped efforts to construct such a proof, and how those efforts were taken up into the public arena.

2. Physics: Simplified Technology and the Laws of Nature

The first independent study of Reagan's proposal came from a group of elite physicists working under the auspices of the Union of Concerned Scientists (Gottfried and Kendall). Panelists included a number of prominent physicists, such as Hans Bethe and Richard Garwin, both long-time advisors within the defense department, and members of the former President's Science Advisory Committee. Their authority rested not only upon their credentials, but also upon their method of analysis. When analyzing the many technological proposals put forward by the Fletcher committee, they appealed directly to technological enthusiasts, assuming that all would "perform as well as the constraints imposed by scientific law permit" (Gottfried and Kendall 2). In this "utopian regime," they used "physics and geometry" to project an element of certainty into the inherently unpredictable future, calculating the cost and effectiveness of various proposals (Gottfried and Kendall 2).

2.1 The Physical Limits of Space-Based Lasers

Consider, for example, their analysis of a proposal to base chemical lasers in space. In a calculation that was to become the subject of considerable controversy, they quickly isolated two primary questions that could be addressed through physical calculations.

First, how many lasers must be stationed on satellites over the Soviet Union in order to shoot down missiles with their rocket boosters, if all are launched at once?

The physicists dubbed this the satellite-booster ratio, noting that it was all a matter of timing: a single laser can only shoot down so many ballistic missiles before they leave the atmosphere. Although the formula for this calculation was determined by physics and geometry, the physicists were forced to make assumptions about future technology, such as the hardness of future missile boosters and the brightness of future lasers. Importantly, the panel assumed impressive technological progress. For example, they assumed that the U.S. would possess lasers ten times as bright as what the Strategic Defense Initiative officially called for. They also assumed that no time would be required to redirect a laser from one target to the next. Using such optimistic assumptions, they concluded that the United States must keep at least 100 lasers over the Soviet Union at all times.

Second, the physicists noted that since the earth spins at one rate and the laser-bearing satellites orbit it at another, the lasers move slightly each time they pass over the Soviet Union. The additional number of satellites that must be set in orbit to account for the spinning earth was dubbed an *absentee ratio*. In their initial calculation, the physicists made the simplifying assumption that orbits were distributed orbits evenly around the earth, giving an absentee ratio of 24. Having determined that 100 lasers must always remain over the Soviet Union, this then required a total of 2,400 satellites.

The physicists used similar methods to analyze many other prospects for space-based defense, such as kinetic-energy hit-to-kill vehicles, proposals to "pop-up" x-ray lasers from earth just in time to intercept missiles, and satellites in geosynchronous orbit. They also analyzed prospects for midcourse and terminal phase defenses. In each phase, they found that the Soviet Union could overcome the defense at less cost than it would take to build it. This led to their most important conclusion: that a space-based missile defense would speed up the arms race on earth and extend it into space.

Working quickly, the physicists released the report just in time for the Senate Armed Committee Hearings on SDI in March of 1984. It was readily picked up by journalists who noted not only the physicists' elite credentials, but also their emphasis on "immutable laws of nature and basic scientific principles." For example, describing the report as a "detailed mathematical analysis," the *New York Times* took note of its quantitative projections, such as the cost of powering lasers in space: 2,400 lasers stationed on satellites would require $70 Billion simply to lift the fuel to the lasers (Mohr, "Study Assails Idea"). An editorial emphasized the point: "a total shield is technologically impossible" (Editorial "The Mirage").

However, within a month of the reports' release, the panel members realized that the satellite calculation was in error. One of the panel directors, physics professor Kurt Gottfried from Cornell, recalled that on a long plane flight after spring break of 1984, he was trying to picture the satellites spinning around the earth and passing over the Soviet Union (Gottfried Interview). He and other members of the panel realized that by choosing orbits carefully, the U.S. might reduce the absentee ratio, and hence the total number of satellites required. When panel member Richard Garwin marched up to Capitol Hill to testify before the Senate Armed Services Committee in April, he noted that this would reduce the total number of satellites to about 800 (Senate 3112). But matters were more complicated still, for the new, more optimal orbits would also shorten the average distance between the satellites and the boosters, reducing the time required to kill a missile. Thus the panel's final, published conclusion was that a space-based defense would require 300 satellites (Bethe et al.).

This revision did not change the physicists' final conclusion: space-based defense could not be expected to change the strategic balance. This was confirmed by a separate, widely-publicized study for the Office of Technology Assessment (OTA). With full knowledge of classified information, physicist Ashton Carter concluded that the prospect of an effective, space-based defense was "so remote that it should not serve as the basis of public expectation or national policy" (Carter 81).

2.2 Speaking with a Measure of Disciplinary Consensus

Despite this growing consensus among physicists, small initial mistakes became a matter of major controversy in the adversarial arena of U.S. politics. Hawks remained committed to investing in defense as well as offense, and Soviet objections to defense did little to curb their zeal. Furthermore, with billions of dollars in research funds at stake, scientists and engineers at the nuclear weapons laboratories were all too eager to publicize the initial error. With the help of sympathetic journalists, these individuals launched an attack on Bethe, Garwin, and others associated with the Union of Concerned Scientists' study.

Conservative pundits soon wrote a series of editorials in the *Wall Street Journal*, accusing these physicists of "politicized" and "shoddy" science (Editorial "Politicized Science"; Editorial "Real Whistle Blowers"; Fossedal). Dartmouth physicist Robert Jastrow launched similar attacks in *Commentary Magazine* ("The War against 'Star Wars'"; "'Star Wars': Robert Jastrow and Critics"). They made their accusations credible by referring to work by Gregory Canavan, a physicist at Los Alamos National Laboratories. Canavan used physics and geometry to demonstrate a so-called "square root law." This suggested, for example, that four Soviet boosters would be necessary to overcome two additional lasers, or that nine Soviet boosters would be necessary to overcome three additional lasers. As leaked to the conservative press, the "square-root law" made missile defense look like a winning game; it would force the Soviet Union to add many missiles to their arsenal in order to overcome just a few lasers.

However, Garwin soon showed how limited a win this would be. In an article published in *Nature*, Garwin showed that the square-root law was only relevant for small numbers of missile boosters. If the Soviet Union were to deploy large numbers of missiles, the satellite-booster ratio grew linearly. Since the Soviets could always punch a hole in the defense by concentrating their missiles at one point, defense was a losing game, at least in economic terms.

In so far as it rested on long accepted laws of nature and geometry, Garwin's calculation could not be challenged. The accusations did not entirely cease in the editorial space of the *Wall Street Journal*, but the physics community began to refer to Garwin's *Nature* article as a standard reference. When the American Physical Society (APS) undertook a review of directed energy weapons for missile defense, the study committee included a version of the calculation in their appendix, reiterating many of Garwin's conclusions (Bloembergen et al.). The committee noted that the previous controversy was "perhaps surprising since . . . the number of satellites required is a problem with a demonstratively correct answer" (Bloembergen et al. 193).

The same was not true of predictions for technological improvement. Physicists at the national laboratories objected to the widely publicized APS conclusion that the feasibility of a defense based on directed energy weapons could not even be evaluated for another decade, and would not be ready to deploy for two decades (Sanger). But as I have discussed elsewhere, the APS effectively backed up Garwin and others associated with the professional society's report (Slayton "Speaking as Scientists"). These and other expressions of consensus among physicists reverberated through the U.S. Congress and national media, marginalizing dissenters, and sending a strong message that exotic weapons would not change the strategic balance anytime soon (see also Slayton "Boycotting Star Wars").

Ongoing objections about even "demonstrably correct" calculations can only be surprising if we forget that the lives of cherished weapons programs rested upon their outcomes. Furthermore, calculations about the feasibility of boost-phase intercept involved disciplined projections into the uncertain world of strategy and politics. The same physical principles could be plied to quite different effect, simply by changing political and strategic assumptions. Nonetheless, the laws of physics enabled the professional physics community to demonstrate certain limits to technology.

However, the process of projecting physical principles into the Star Wars debate tended to marginalize other important aspects of the defensive system. The physicists recognized that their utopian projection assumed "that the battle management software is never in error," despite the defense department's own acknowledgement that a space-based missile defense system would "face formidable systemic problems" (Gottfried and Kendall 2; Bethe et al. 48). Physics did not encourage deep analysis of the challenges of complex, humanly-built systems. As we will see below, this analysis was facilitated instead by the discipline of software engineering.

3. Computing: Complex Systems and Engineering Experience

Significantly, the Fletcher panel on "Battle Management, Communications, and Data Processing" focused on managing the complexity of a global missile defense system. While acknowledging the challenges of "achieving the computational speed and capacity needed to make decisions and to manage a complex and rapidly evolving battle," their more urgent concern involved "the difficulty of specifying and designing a system that will be of unprecedented complexity" and "the reliability and safety of any resulting system that may finally be deployed" (McMillan et al. 4).

As the panel discussed, software would be needed for battle management at every level of command and in each stage of defense. In each local system, software-hardware systems would need to track objects, distinguish threatening objects from decoys, chaff and false targets, allocate weapons to threatening objects, and assess the success of a "kill." Global battle management would entail perpetual threat surveillance, managing rules of engagement, assessing the battle status, coordinating the operation of sub-systems, and managing resources so that the defense itself would be able to survive. Though humans would not entirely be shut out of the decision making loop, very short time-scales would require an unprecedented level of automation (McMillan et al. 25).

While "granting that rules of engagement can be designed that are considered both safe and effective," the panel went on to note that it would be difficult to automate them "in such a way that one is confident that they will work as designed" (McMillan et al. 27). The battle management software would be between three and five times larger than any other software system ever developed, up to ten million lines of code. It would also need to be flexible, adapting to changing strategies and Soviet countermeasures. The tremendous system complexity, combined with lack of opportunities for operational testing, presented unprecedented challenges for ensuring reliability.

Thus, the panel's first conclusion was: "Specifying, generating, testing and maintaining the software for a battle management system will be a task that far exceeds the complexity and difficulty that has yet been accomplished in the production of civil or military software systems," concluding: "considerably more emphasis must be placed on this area to insure that the proper technology will be available to develop a BMD battle management system" (McMillan et al. 4, 46).

Appealing to the need for technical progress, the panel stopped short of arguing that it would be impossible to produce trustworthy software for a Star Wars system. But when independent scientists and engineers sized up the problem, this is precisely what many concluded. Once again, countermeasures were the central problem. Without fully knowing the countermeasures that a system would face, how could the battle management software be adequately specified, let alone tested? Scientists who were convinced that the SDI software could not be trustworthy faced a challenge: to project this conclusion into the Star Wars debate in a compelling way.

3.1. The Need to Test Battle Management Software

Significantly, they did so by turning not to laws of nature, but to engineering experience. For example, Herbert Lin, a young physicist at MIT, used formulae in the textbook *Software Engineering Economics* (Boehm), to calculate that 81,700 programmer-years would be required to produce 10 million lines of code. In an article published in *Scientific American*, Lin outlined four major steps in software development – planning, design, implementation, and testing – showing that significant challenges could be anticipated in each of these steps (Lin). Most notably, he used well-documented system failures to argue that inadequate specifications and hidden flaws would only be discovered and corrected with operational experience. Yet, he argued, "a comprehensive ballistic-missile defense requires not only that software operate properly the first time, in an unpredictable environment and without large-scale empirical testing, but also that planners are positive it will do so" (Lin 53).

As this suggests, computing directed attention to a very different set of concerns than those of physics. Where physics encouraged analysts to project simplified, perfect systems, subject only to the constraints of nature, computing suggested complex, glitch-ridden systems, subject to social and organizational constraints. We can see these differences visually in the distinctive diagrams associated with the senior physicists' study and Lin's study, as published in *Scientific American*. While the senior physicists' projection of a global defense system highlighted physical arrangements in a deliberately simplified fashion (Figure 1), Lin's image of a global defense system highlighted the complexity of information, abstracting from the physical world to emphasize decisions (Figure 2).

Though all of these critiques were set in motion in the days and weeks following Reagan's speech, it was not until the summer of 1985 that SDI software began to attract major public attention. In July of 1985, the *New York Times* revealed that a prominent software engineer by the name of David Parnas had resigned from a defense department panel on computing for missile defense (Mohr "Scientist Quits"). Parnas drew attention in part because he was the first scientist known to formally resign from a government panel, and he repeatedly emphasized conflicts of interest in the advisory system, including the panel he resigned from (subsequently named the Eastport panel).

However, he also drew attention for his specific critique of the SDI software, which went much further than previous analysts. Like others, he emphasized engineering experience, including his own twenty years of experience and prominence in the field of software engineering: "I am willing to stake my professional reputation on my conclusions" (Parnas Letter 1). But Parnas grounded his critique in more than individual or collective engineering experience. In a series of eight short papers appended to his letter, subsequently published and known as the "octet," Parnas argued that SDI software would not be reliable because of a "fundamental difference" between traditional engineering products and software systems (Parnas "Software Aspects"). He cited the convergence of two features of software. First, software products were discrete state systems, not governed by well understood mathematical functions, and therefore more unpredictable. And second, they were too large to be exhaustively tested. Parnas argued that these two features combined to make software systems inherently less reliable than traditional engineering products, calling this "a fundamental difference that will not disappear with improved technology" (Parnas, "Software Aspects" 1328).

Significantly, by claiming that fundamental differences would limit technical progress, Parnas was trying to argue that software risks were not a *social* problem associated with human engineering, but a *natural* problem of complexity, one that human ingenuity could not circumvent. Were such fundamental differences accepted by the software engineering community, he could have projected an element of disciplinary objectivity into the debate, arguing for limits to technological progress.

3.2 Disciplinary Dissent: Fundamental Limits to Software?

However, matters were not so simple. As I have discussed elsewhere, many in the computing community contested Parnas' claim about "fundamental" limits (Slayton "Speaking as Scientists"). For example, Danny Cohen, the chairman of the Eastport panel, repeatedly emphasized: "There is no fundamental result saying that it is impossible to meet the SDI computing requirements." His audience was not always very receptive to his remarks. In a debate at MIT, one individual objected: "You said building SDI does not contradict any fundamental law . . . . It does contradict Murphy's law" (quoted in Jacky).

Nonetheless, Cohen and others continued objecting that there were no fundamental laws that made it impossible to develop trustworthy Star Wars software. As Parnas went to great pains to point out, he not claimed that Star Wars was impossible, merely improbable. As he explained in a 1985 debate:

> It's not impossible that this thing will be right the first time, it's not impossible that if you take 10,000 monkeys and let them type for 5 years that they'll recreate the Encyclopedia Britannica. It's possible. I wouldn't count on it, in either case . . .
>
> . . . .
>
> Is there any fundamental reason why it's unlikely to work correctly when really needed? Yes, because we don't know what it's supposed to do, and we don't have any way to verify that it does what we think it should . . . . We have lots of experience in building software, we know what human beings can do and they can't build software that's right before it's really used. And we know good mathematical reasons why that's going to be the case. (Goldberger 26)

As this suggests, Parnas felt pressure to give a mathematical explanation for Murphy's Law – that all-too-familiar axiom that if something can go wrong, it will. Yet notably, he repeatedly returned to engineering experience rather than fundamental laws. To many in the computing community, this seemed to describe socially contingent phenomena rather than absolute limits.

Significantly, the Eastport panel's final report did acknowledge a problem with the software development for SDI. But this was not an unchangeable, "natural" problem; it was a "social" problem. In particular, they claimed that the SDI systems contractors had ignored the advice of the Fletcher report two years earlier, that "the battle management system and its software must be designed as an integral part of the ballistic missile defense (BMD) system as a whole, not as an appliqué". Instead, the companies had "treated the battle management computing resources and software as a part of the system that could be easily and hastily added," and "as something that is expected to represent less than five percent of the total cost of the system, and therefore could not significantly affect the system architecture." They accused the contractors of developing "their proposed architectures around the sensors and weapons and have paid only lip service to the structure of the software that must control and coordinate the entire system" (Cohen 10).

Nonetheless, the panel concluded that the computer system was achievable, if the program was *managed* in a way that recognized computing as the "paramount" problem of missile defense. In particular, they argued that a "loosely coupled" system could be made reliable. "The idea is simple," they explained. "Test each independent platform or group separately. If it works, then its independence allows one to infer that the whole will work also" (Cohen 20-21).

The computing debate did not end with the Eastport study report. Many in the computing community were quick to object that it would be impossible to sufficiently decouple systems components, and that even loose coupling would likely result in unexpected interactions, interactions that could only be fully debugged through testing. The professional computing community never reached consensus on the issues surrounding SDI software. When Parnas' critique of SDI was published in the *Communications of the ACM* (Parnas "Software Aspects"), responses were divided. Among ten total published responses, five disagreed that SDI computing was hopeless, two disagreed with portions of Parnas' argument, and only three expressed total agreement. Several respondents emphasized that human ingenuity or technological progress would make it possible to develop reliable software. As one insisted, "with reasonable progress in the state of the art, this is achievable. At the very least, there has not been adequate proof that it is not achievable" (VerHoef and Bunting).

As this suggests, many in the community felt that engineering experience was not sufficient proof of the unreliability of SDI software. Technological progress is perhaps *the* central vocation of computer engineers, and many felt that any statement of fundamental limits undermined this vocation. Thus, it is perhaps unsurprising that professional computing associations never conducted a study of software for a global missile defense system. Unlike the American Physical Society, they never projected their disciplinary objectivity into the SDI debate.

4. Conclusion: Proof and Common Sense in the Star Wars Debate

To summarize: physics and software engineering both offered scientists and engineers ways of projecting an element of certainty into the complex and uncertain future of a "Star Wars" missile defense system. However, these projections differed in ways that suggest distinctive disciplinary cultures, and that generated very different kinds of authority in the Star Wars debate.

Physics enabled scientists and engineers to project limits on deliberately simplified, idealized technology. By focusing on the constraints of nature rather than human beings, on representing the immutable physical world rather than intervening in the social world, professional physics associations were able to gain broad professional consensus. By contrast, the projections of software engineering directed attention to the challenges of developing and maintaining complex, humanly designed systems. In so far as it involved management, human agents were intrinsically embedded in the quantitative projections of the computer experts. Software engineers felt a vocational commitment to improvement, and the computing profession was not able to generate consensus about fundamental limits to software technology.

But which of the challenges—those of complex software or those of high tech weapons—proved more persuasive for policymakers? Taken as a solitary debate, the Star Wars debate yields no clear verdict. Despite the lack of consensus within professional computing associations, the controversy surrounding Parnas' resignation brought the issue to the attention of policymakers. In 1986, the U.S. Congress commissioned its Office of Technology Assessment (OTA) to study the "Star Wars" software problem. Two years later, the *Washington Post* was among major newspapers to report the findings: "President Reagan's proposed missile defense system likely would 'suffer a catastrophic failure' the first—and therefore only—time it was used to protect the United States against a Soviet nuclear attack" (Smith).

Significantly, this report did not go uncontested. OTA Director John Gibbons and the director of the Strategic Defense Initiative Organization (SDIO) James Abrahamson, soon wrote a letter taking "serious issue" with the story:

> SDIO and OTA disagree on the feasibility of reliable, trustworthy software for a future defense against ballistic missiles. SDIO believes that such software can be developed; OTA is much more skeptical, and contends that there would be a 'significant probability' of 'catastrophic failure' of a ballistic missile defense system resulting from a software error. Nowhere in its study, however, does OTA conclude—as alleged by *The Post*—that such a failure would be 'likely.' As any statistician knows, and as the OTA report should make clear, 'significant probability' does not equate to likelihood. (Abrahamson and Gibbons)

It is tempting to dismiss such distinctions, between "significant probability" and "likely" failure, as mere semantic quibbling. But this would be a mistake.

Instead, I argue that this exchange reveals much about American political culture and the specific forms of authority that computing holds within that culture. The *Washington Post* article suggests that the computing revealed the limits of "Star Wars" much more persuasively than did physics, by grounding its arguments in common sense. Familiar with Murphy's Law, laypersons were ready and willing to take home the message that Star Wars was "likely" to fail catastrophically – and for all practical purposes, this was the more accurate message. For statistics cannot fully capture the deeper questions of trustworthiness in a "Star Wars" system. When the consequences of catastrophic failure involve nuclear devastation, any rational policymaker must consider a "significant probability" to be "likely." Common sense recommends gambling with such distinctions in the face of even one nuclear bomb. While arguments about software reliability were quite accessible, the same cannot be said of the physicists' debate. Without the intervention of the APS, it would have been exceedingly difficult for laypersons to discern an objective position in the debate about orbiting satellites.

Yet the distinction between "significant probability" and "likely" also suggests the cause of the computing community's inability to reach consensus on Star Wars software, either through formal disciplinary means, or through government panels: a faith in technological progress. For a discipline committed to technological progress, consensus about the unreliability of software entailed not only proving improbability, but impossibility.

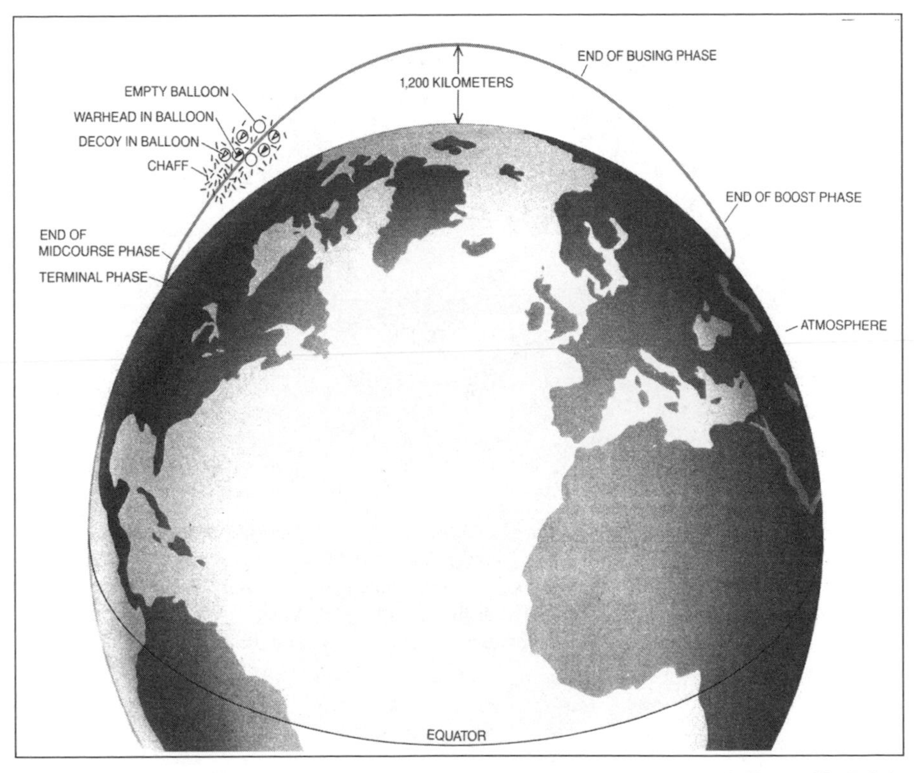

© George V. Kelvin

Figure 1: The physical challenges for a global missile defense. An Intercontinental Ballistic Missile (ICBM) undergoes four phases of flight. The optimum time to shoot down a missile is in its boost phase, as it launches out of the earth's atmosphere. In 1984, this period lasted three to five minutes, but engineers had devised ways of shortening it to one minute. In the "busing" phase, each booster could release up to ten Multiple Independent Re-entry Vehicles (MIRVs), containing warheads heading toward different targets. They might also release of lightweight decoys, chaff, and other deceptive devices to hide the missile during the remainder of the midcourse phase. During the final few moments of terminal phase, lightweight decoys and chaff would be slowed by the earth's atmosphere, and the real warheads would emerge speeding toward their targets (Bethe 40).

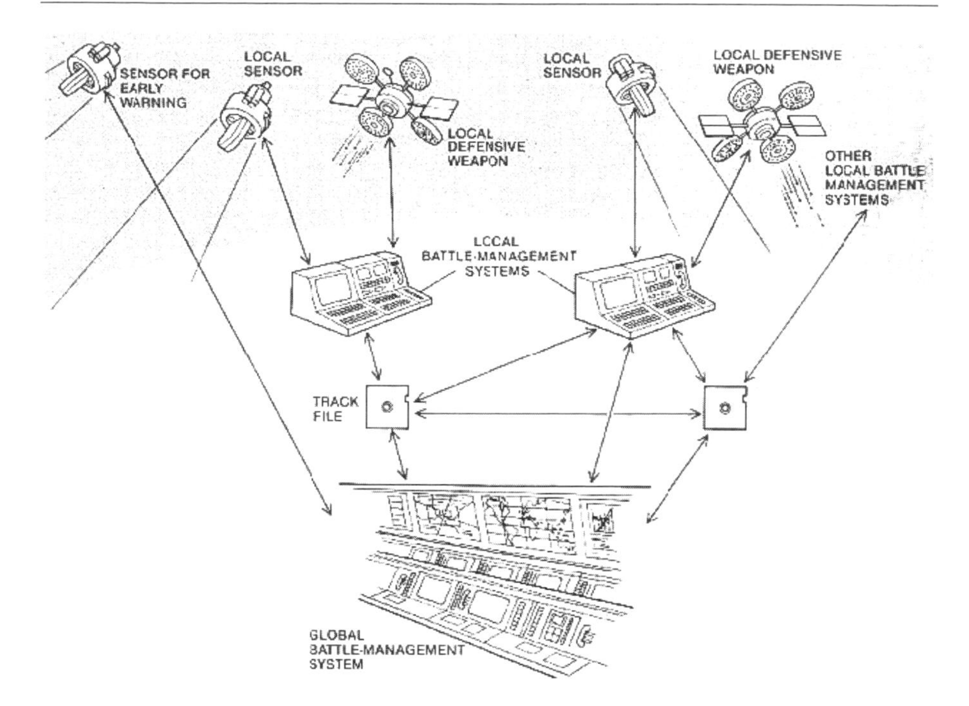

Figure 2. The computing challenges for global missile defense. Because of the very short timescales involved, the missile defense battle management would require a high level of automation. Software would be required for managing sensor data, controlling individual weapons, local battle management, and the overall tracking and coordination of a global battle. The software would be three to five times larger and more complex than any built to date, but there would be limited opportunities for realistic testing (Lin 50).

Works Cited

Abrahamson, James A. and John H Gibbons. "That Sdi Story." *Washington Post* 5 May (1988).

Barber, Richard. *History of Darpa*. Alexandria: Institute for Defense Analysis, 1975.

Bethe, Hans, et al. "Space-Based Ballistic-Missile Defense." *Scientific American* 251.4 (1984): 39.

Bloembergen, N, et al. "Aps Study: Science and Technology of Directed Energy Weapons. " *Rev Mod Phys* 59.3 (1987): S3.

Boehm, Barry. *Software Engineering Economics. Prentice-Hall Advances in Computing Science and Technology Series*. Englewood Cliffs: Prentice-Hall, 1981.

Broad, William. *Teller's War: The Top Secret Story Behind the Star Wars Deception*. New York: Simon & Schuster, 1992.

Cannon, Lou. "President Seeks Futuristic Defense against Missiles." *Washington Post* 24 March 1983: A1.

---. "Reagan-Gorbachev Summit Talks Collapse as Deadlock on Sdi Wipes out Other Gains." *Washington Post* 13 October 1986: A1.

Carter, Ashton. *Directed Energy Missile Defense in Space: Background Paper*. Washington, D.C.: The Office of Technology Assessment, 1984.

Cohen, Daniel. *Eastport Study Group: A Report to the Director of the Strategic Defense Initiative Organization*: SDIO, 1985.

Downer, John. "When the Chick Hits the Fan: Representativeness and Reproducibility in Technological Tests." *Social Studies of Science* 37.1 (2007): 7-26.

Downey, Gary L. and Juan C. Lucena. "Knowledge and Professional Identity in Engineering: Code-Switching and the Metrics of Progress." *History and Technology* 20.4 (2004): 393-420.

Editorial. "The Mirage of Space Defense." *New York Times* 2 April 1984: A18.

---. "Politicized Science." *Wall Street Journal* 10 December 1984: 26.

---. "Real Whistle Blowers." *Wall Street Journal* 5 April 1985: 16.

Epstein, Steven. *Impure Science: Aids, Activism, and the Politics of Knowledge*. Los Angeles, CA: University of California Press, 1996.

Fitzgerald, Francis. *Way out There in the Blue: Reagan, Star Wars, and the End of the Cold War*. New York: Simon & Schuster, 2000.

Fossedal, Gregory A. "Star Wars and the Scientists." *The Wall Street Journal* 16 June 1985.

Galison, Peter. *Image & Logic: A Material Culture of Microphysics*. Chicago, IL: University of Chicago Press, 1997.

Garwin, Richard L. "How Many Orbiting Lasers for Boost-Phase Intercept?" *Nature* 315 (1985): 286-290.

Goldberger, Marvin (Debate Moderator). SDI Debate. Stanford University, CA: Stanford Computer Science Department, 1985.

Gottfried, Kurt and Henry W Kendall. *Space-Based Missile Defense: A Report by the Union of Concerned Scientists*. Cambridge: Union of Concerned Scientists, 1984.

---. Phone Interview. 12 July 2003.

Gwertzman, Bernard. "Reagan-Gorbachev Talks End in Stalemate as U.S. Rejects Demand to Curb 'Star Wars'." *New York Times* 13 October 1986: A1.

Herken, Gregg. *Cardinal Choices: Presidential Science Advising from the Atomic Bomb to SDI*. New York: Oxford University Press, 1992.

Hiatt, Fred. "Limited Abm Is Urged to Protect U.S. Missiles." *The Washington Post* 8 March 1984: A30.

---. "Reagan's 'Star Wars' Uncertain after Year." *The Washington Post* 24 March 1984: A5.

Hilgartner, Stephen. *Science on Stage: Expert Advice as Public Drama*. Stanford, CA: Stanford University Press, 2000.

Hughes, Thomas. *American Genesis*. New York: Penguin Books, 1989.

Jacky, Jon. "Cpsr/Boston Co-Sponsors Debate on Computer Requirements of Star Wars." *The CPSR Newsletter* Fall 1985: 1.

Jasanoff, Sheila. *The Fifth Branch: Science Advisors as Policymakers*. Cambridge, MA: Harvard University Press, 1990.

---. "Science, Politics, and the Renegotiation of Expertise at Epa." *Osiris* (1992): 194-217.

---. *Designs on Nature: Science and Democracy in Europe and the United States*. Princeton, NJ: Princeton University Press, 2005.

Jastrow, Robert. "'Star Wars'; Robert Jastrow and Critics." *Commentary* June (1984): 7-12.

---. "The War against 'Star Wars'." *Commentary* December (1984): 19-25.

Kaiser, David. *Drawing Theories Apart: The Dispersion of Feynman Diagrams in Postwar Physics*. Chicago, IL: University of Chicago Press, 2005.

Knorr-Cetina, Karin. *Epistemic Cultures: How the Sciences Make Knowledge*. Cambridge: Harvard University Press, 1999.

Kuhn, Thomas S. *The Structure of Scientific Revolutions*. Chicago, IL: University of Chicago Press, 1962.

Lin, Herbert. "The Development of Software for Ballistic-Missile Defense." *Scientific American* 253.6 (1985): 46-53.

McMillan, Brockway et al. *Eliminating the Threat Posed by Nuclear Ballistic Missiles, Vol. V: Battle Management, Communications, and Data Processing*. Washington, D.C., 1984.

Mohr, Charles. "Scientist Quits Antimissile Panel, Saying Task Is Impossible." *New York Times* 12 July 1985: A6.

---. "Study Assails Idea of Missile Defense." *New York Times* 22 March 1984: A11.

Moore, Kelly. *Disrupting Science: Social Movements, American Scientists, and the Politics of the Military, 1945-1975*. Princeton, NJ: Princeton University Press, 2008.

Parnas, David. Letter to James H. Offut, Assistant Director, BM/C3, Strategic Defense Initiative Organization. 28 June 1985.

---. "Software Aspects of Strategic Defense Systems." *Communications of the ACM* 28.12 (1985): 1326.

Porter, Theodore. *Trust in Numbers*. Princeton, NJ: Princeton University Press, 1996.

Pratt, Eric. *Selling Strategic Defense: Interests, Ideologies, and the Arms Race*. Boulder: Lynne Rienner Publishers, 1990.

Robinson, Clarence A. "Panel Urges Defense Technology Advances." *Aviation Week & Space Technology* 17.10 (1983): 16.

---. "Study Urges Exploiting Technologies." *Aviation Week & Space Technology* 24.10 (1983): 50.

---. "U.S. Strategic Defense Options; Panel Urges Boost Phase Intercepts." *Aviation Week & Space Technology* 5.12 (1983): 50.

Sanger, David E. "Missile Defense: New Turn in Debate." *New York Times* 24 April 1987: A8.

---. "Speaking as Scientists: Computer Professionals in the Star Wars Debate." *History and Technology* 19.4 (2004): 335-364.

Slayton, Rebecca. "Discursive Choices: Boycotting Star Wars between Science and Politics." *Social Studies of Science* 37.1 (2007): 27.

Sluijs, Jeroen van der et al. "Anchoring Devices in Science for Policy: The Case of Consensus around Climate Sensitivity." *Social Studies of Science* 28.2 (1998): 291-333.

Smith, Jeffrey. "SDI Faulted in 2-Year Hill Study; 'Catastrophic Failure' Likely If Ever Used, Scientific Group Says." *Washington Post* 24 April 1988: A1.

Snow, C. P. *The Two Cultures and the Scientific Revolution*. The Rede Lecture. Cambridge, 1959.

Traweek, Sharon. *Beamtimes and Lifetimes: The World of High Energy Physicists*. Cambridge, MA: Harvard University Press, 1992.

United States. Cong. Senate. Committee on the Armed Services, Subcommittee on Strategic and Theater Nuclear Forces. *Department of Defense Authorization for Appropriations for Fiscal Year 1985. Part 6. The Strategic Defense Initiative: Defensive Technologies Study*. Hearing, 8, 22, March, 24 Apr. 98th Cong., 2nd sess. Washington, D.C.: Government Printing Office, 1984.

VerHoef, Edward W. and David C. Bunting. "Still More on Software Aspects of Strategic Defense Systems." *Communications of the ACM* 29.9 (1986): 830-831.

Notes on Contributors

Klaus Benesch is Professor of English and American Studies at the University of Munich and Director of the Bavarian American Academy (Munich). He was a 2004 Mellon Fellow at the Harry Ransom Humanities Research Center of the University of Texas (Austin), and has taught at the University of Massachusetts (Amherst) and Weber State University (Utah). Previous publications include: *The Power and Politics of the Aesthetic in American Culture* (editor/2007); *African Diasporas in the New and Old Worlds: Consciousness and Imagination* (editor/2006/04); Space *in America: Theory, History, Culture* (editor/2005); *The Sea and the American* Imagination (editor/2004); and *Romantic Cyborgs: Authorship and Technology in the American Renaissance* (2002).

Hanjo Berressem teaches American Literature and Culture at the University of Cologne. His interests include literary theory, contemporary American fiction, media studies, and the interfaces of art and science. Apart from articles situated mainly in these fields, he has published books on Thomas Pynchon (*Pynchon's Poetics: Interfacing Theory and Text*, 1992) and on Witold Gombrowicz (*Lines of Desire: Reading Gombrowicz's Fiction with Lacan*, 1998). He is the editor, together with D. Buchwald und H. Volkening, of the collection *Grenzüberschreibungen: Feminismus und Cultural Studies* (2001), and, together with D. Buchwald, of *Chaos-Control / Complexity: Chaos Theory and the Human Sciences* (2000).

Robin Morris Collin currently teaches law on the faculty of Willamette University College of Law which she joined in 2003 after a distinguished career teaching law at the University of Oregon, Washington & Lee College of Law, and Tulane University among other schools. She and her husband, Robert W. Collin, have published numerous works exploring the themes of the Environmental Justice Movement in the United States, and the International Movement Toward Sustainability. These include "Forever Wild, Forever Free: Sustainability and Equity," in: *From Landscape to Technoscape: Concepts of Space in American Culture* (2005); "Environmental Reparations for Justice and Sustainability," in: *The Quest for Environmental Justice Human Rights and the Politics of Pollution* (2005); "The Role of

Communities in Environmental Decisions: Communities Speaking for Themselves," in: *Journal of Environmental Law And Litigation* 13 (1998); and "Where Did All the Blue Skies Go: Sustainability and Equity: The New Paradigm," in: *Journal of Environmental Law and Litigation* 9 (1994). She was the first professor of law to teach a course on *Sustainability in Law and Policy in the United States*, and she was awarded the David Brower Lifetime Achievement Award for Environmental Activism in 2002.

Peter Freese is Professor emeritus of American Studies at Paderborn University. He was visiting professor at universities in Britain, the U.S. and Hungary and Fellow in Residence at the Claremont Colleges, holds honorary doctorates from Lock Haven University of Pennsylvania and Dortmund University, and is the recipient of the 'Bundesverdienstkreuz am Bande'. Among his 40 books are *Die Initiationsreise* (1971; rpt. 1998); *'America': Dream or Nightmare?* (1990; 3rd rev. edition, 1994); *The Ethnic Detective* (1992); *From Apocalypse to Entropy and Beyond* (1997); and *Teaching 'America': Selected Essays* (2002). He has contributed 180 articles to journals and anthologies and is the editor of the monograph series *Arbeiten zur Amerikanistik* and of the *VIEWFINDER* series for the advanced EFL-classroom. He is currently working on a monograph on Kurt Vonnegut's novels.

Ursula K. Heise is Associate Professor of English & Comparative Literature at Stanford University, where she teaches contemporary literature and literary theory. Her areas of research and teaching include 20th century fiction and poetry, ecocriticism, literature and science, literature and new media, and theories of modernization and globalization. She is the author of *Chronoschisms: Time, Narrative, Postmodernism* (1997) and *Sense of Place and Sense of Planet: The Environmental Imagination of the Global* (2008). She is currently working on a book project entitled *The Avantgarde and the Forms of Nature*.

Heike Mayer is Assistant Professor in the Urban Affairs and Planning program at Virginia Tech's Alexandria Center. She studied at the University of Konstanz and received her master's degree and Ph.D. in Urban Studies from Portland State University. Her doctoral work focused on the evolution of Portland's high-technology industry—also known as Silicon Forest—in the absence of a major, world class research university such as MIT or Stanford. Her research interests are in regional economic development, high-technology regions, entrepreneurship and innovation. She is currently working on a variety of research projects. One examines the evolution of second-tier high-tech regions in the absence of world-class universities. The second project focuses on women high-tech entrepreneurs in Silicon Valley, Boston, Portland, and Washington, D.C. She is also working on a study of Slow Cities in Europe (with Paul Knox). Her work has been published in the *Journal of the*

American Planning Association; *Journal of Urban Affairs*; *Economic Development Quarterly*; *Economic Development Journal*; and by the *Brookings Institution*.

Suzanne Nalbantian is Professor of Comparative Literature, at Long Island University and an interdisciplinary scholar. Her specialization is in modern Western literature, memory study, and links to neuroscience. Her books include *Memory in Literature: From Rousseau to Neuroscience* (2003); *Aesthetic Autobiography: From Life to Art in Marcel Proust, James Joyce, Virginia Woolf, and Anaïs Nin* (1994); *Anaïs Nin: Literary Perspectives* (1997); *Seeds of Decadence in the Late Nineteenth-Century Novel* (1983); and *The Symbol of the Soul from Hölderlin to Yeats* (1977). She holds a Ph.D. from Columbia University and is a permanent member of Columbia's Society of Fellows in the Humanities. She is the winner of the TASA Award for Lifetime Scholarly Achievement at Long Island University. She has been the recipient of a National Endowment for the Humanities Fellowship for Independent Study and Research. She has lectured widely on the subject of memory at universities in the U. S. and Europe, including Yale, Stanford, Columbia, Indiana, Carnegie-Mellon, University of Pittsburgh, and Sorbonne-Paris IV as well as, most recently, at labs such as the Cold Spring Harbor Laboratory on Long Island, the Max-Planck Institute in Tübingen, Germany, and the Collège de France and the Pasteur Institute in Paris.

Alondra Nelson is Assistant Professor of African American Studies, American Studies, and Sociology at Yale University. She is author of *Body and Soul: The Black Panther Party and the Politics of Health and Race* (forthcoming) and co-editor of *Technicolor: Race, Technology, and Everyday Life.* Her research centers on historical and socio-cultural studies science, technology, and medicine with an emphasis on racial formation processes, social movements, and identity. This chapter is adapted from the previously published article "Bio Science: Genetic Genealogy Testing and the Pursuit of African Ancestry," *Social Studies of Science* 38 (October 2008): 759-783.

David E. Nye is the author or editor of 18 books on American history and society, most of which deal with technology and culture. His most recent book is *Technology Matters; Questions to Live With* (2006) [German and French translations appeared in 2007]. In 2005 he received the Leonardo da Vinci Medal from the Society for the History of Technology, the highest award of that organization. He is Professor of American Studies at the University of Southern Denmark.

Denise Pilato received her doctorate in American Studies from Michigan State University in 1998. She is an Associate Professor in the College of Technology at Eastern Michigan University where she teaches in the Master of Science Program and Ph.D. Program in Technology Studies. She teaches interdisciplinary courses in American culture with a central focus on the social impact of science and technology. Her research interests address issues related to gender and technology, with a specific interest in 19th century and early 20th century women inventors. Scholarly work on historic women inventors has been a focus of much of her primary research, and include the publication of her book, *The Retrieval of a Legacy: Nineteenth Century Women Inventors* (Praeger Publishers 2000), which places the experiences of women inventors in an historical and cultural context in 19th century America. Her work on women inventors also appears in the *International Journal of Naval History,* Vol. 1, April 2000, "Martha Coston: A Woman, a War, and a Signal to the World," and a forthcoming article in *Minerva: Women & War*, Spring, 2009 entitled, "The Use of Coston Flares by the U.S. Navy in Civil War Blockade Operations: "A Powerful Auxiliary of Incalculable Value." She has presented her work at numerous national and international conferences.

Rebecca Slayton is a lecturer and researcher in the Science, Technology, and Society Program and an affiliate in the Center for International Security and Cooperation, both at Stanford University. She earned her doctorate in chemistry at Harvard in 2002 before retooling as a postdoctoral fellow in the Science, Technology, and Society Program at the Massachusetts Institute of Technology (MIT). Her research focuses upon public constructions of science, technology, and international security. She is writing a book about how science and engineering disciplines have shaped expert advising and debate on missile defense in the United States.

Joseph Tabbi is the author of *Cognitive Fictions* (2002) and *Postmodern Sublime* (1995), books that examine the effects of new technologies on contemporary American fiction. He edits the electronic book review and has edited and introduced William Gaddis's last fiction and collected non-fiction. His essay, "The Processual Page," appears in the *Journal of New Media and Culture*. Also online (the Iowa Review Web) is an essay-narrative, titled "Overwriting," an interview, and a review of his recent work. He is a director of the Electronic Literature Organization and Professor of English at the University of Illinois at Chicago.